Routledge Revivals

British Locomotive Design

First published in 1967, *British Locomotive Design* is a review of British locomotive practice during the century and a quarter of steam's ascendency. The author shows how the Stephenson basic plan, while remaining essentially unchanged, was adapted over the years to meet the increasing demands made upon it in terms of haulage power and speed. The subject's widespread appeal is derived largely from the enormous variety of practical examples furnished by the products of the former companies, the groups and nationalisation, and, in the description of these by one who has a lifelong predilection for the steam engine in all its manifestations, readers will find due recognition accorded to the machines which had won their favour and, with others, have exercised a fascination as unquestionable as it is difficult to account for in set terms. This book will be of interest to students of engineering, technology, design and history.

British Locomotive Design

1825-1960

Graham Glover

First published in 1967
By George Allen & Unwin Ltd

This edition first published in 2023 by Routledge
4 Park Square, Milton Park, Abingdon, Oxon, OX14 4RN
and by Routledge
605 Third Avenue, New York, NY 10017

Routledge is an imprint of the Taylor & Francis Group, an informa business

Publisher's Note
The publisher has gone to great lengths to ensure the quality of this reprint but points out that some imperfections in the original copies may be apparent.

Disclaimer
The publisher has made every effort to trace copyright holders and welcomes correspondence from those they have been unable to contact.

A Library of Congress record exists under LCCN: 67112522

ISBN: 978-1-032-62922-3 (hbk)
ISBN: 978-1-032-62924-7 (ebk)
ISBN: 978-1-032-62923-0 (pbk)

Book DOI 10.4324/9781032629247

BRITISH LOCOMOTIVE DESIGN

1825–1960

GRAHAM GLOVER

WITH PHOTOGRAPHS SELECTED AND SUPPLIED
BY J. H. COURT

London

GEORGE ALLEN AND UNWIN LTD

RUSKIN HOUSE MUSEUM STREET

FIRST PUBLISHED IN 1967

PRINTED IN GREAT BRITAIN
in 11 point Plantin type
BY C. TINLING AND CO. LTD
PRESCOT

PREFACE

So many books have been written in recent years about the steam locomotive that a very good reason is required to justify an addition to their number.

The present writer has sought to tackle the subject from a new angle. His concern is with basic design or, if the phrase may be allowed, with the architecture of the locomotive. The purpose is to show, against the background of essential principles, how for a century and a quarter locomotive engineers have adapted the Stephenson ground-plan to the constantly increasing demands made upon it by the exigencies of traffic in terms of haulage power and speed.

Notwithstanding the great advances in research made in recent years, the steam engine remains in some respects an enigma. It is pleasant to think that it has passed from the scene without revealing all its secrets. One is reminded of the story told of one of the most recent classes of locomotive, which would not steam satisfactorily until the blast arrangements of a successful class produced more than 50 years earlier, had been studied and incorporated.

An enigma! Quite a number of classes have never fulfilled the promise of success which their excellent proportions seemed to warrant. Others of cramped design have performed excellently. Every ground-plan has its limits. The shoe will pinch somewhere and, with increasing girth, more painfully. Certain designers in the course of locomotive history have cut clearances to a minimum and exploited to the utmost their chosen plan; others have adopted more easy going methods; others, again, seem to have been largely unaware of inherent obstacles to efficient operation or to have taken little pains to eliminate them. General design affords an absorbing study which, in the voluminous literature devoted to the locomotive, has too often been neglected or insufficiently regarded. The present writer has sought in some measure to repair this omission.

Products of human ingenuity have their alloted span. From crude beginnings they achieve stability in a recognizable form and flourish for a time, only to fall a victim to that self-same ingenuity which gave them birth. So it has proved with the subject of this volume. The steam locomotive has had a good innings. If the end has come, it is useless to repine. At least an unique occasion is presented for a review of its entire career.

The Stephenson engine, with its tubular boiler barrel, bounded at one end by a water-jacketed firebox and at the other by a sealed-off chamber through which the hot gases from the fire and the exhaust steam from the cylinders are discharged into the atmosphere, was evolved within a few years of the appearance of the first engine to run on the first public railway—*Locomotion*, of 1825. The rest of the period, down to the appearance of the most advanced product of British Railways—the last of which was, appropriately enough, called *Evening Star*—witnessed the progressive development (but no more) of this basic plan.

The plan itself imposed its own inherent limitations. But there were others also. Designers had to work within restrictions imposed by the exacting British loading gauge and the weight capable of being carried by track and underline bridges. Much of the interest of our study lies in observing the various ways in which these difficulties were surmounted.

The central portion of the book is founded on a series of articles on British Locomotive Design which have been appearing in the *Railway Magazine* over the past few years, and the writer thanks the proprietors and editor of that publication for permission to reproduce this material. These articles covered the period 1890–1948. Additional chapters complete the picture.

The general division of the subject is dictated by well defined periods of evolution. The first chapter, 'Beginnings', describes the developments which took place from the crude originals down to the end of the '50s, by which time the locomotive had attained, in all essentials, its modern form. 'Mid-Victorian' (chapter 2) spans thirty years of a process of refinement of the basic model. 'Late Victoriana' (chapter 3) and 'Edwardian Engines' (chapter 5)—with 'The Broad Gauge' (chapter 4) interposed—together cover a period of some twenty years, during which important changes, both in form and size, were introduced in response to the demand for machines of greater power. 'The Last Years' (chapter 6) witnessed the final stage of the great variety of types produced by the former companies for virtually similar requirements and (four years of war notwithstanding) was one of the most colourful periods in the annals of British locomotive history. 'Grouping' (Chapter 7), which followed in 1923, inevitably entailed the loss of much cherished work, but produced highlights of speed and haulage power never before achieved. Finally (chapter 8) with nationalization in 1948, we encounter a series of new designs produced to meet almost the entire range of British locomotive requirements. These periods vary considerably in duration, as do also the lengths of the corresponding chapters. But the division is logical enough and provides convenient vantage points from which to view the subject as a whole.

Considerations of space have demanded a rigid economy in the selection of the great body of material available. A number of important classes have necessarily been passed over or accorded only the briefest mention. But it is hoped that no developments of real significance have been omitted and that all the multitudinous classes described have a direct bearing on the main theme—progress in design.

One aspect of the steam locomotive which has been only lightly touched upon in the text is that of appearance. Any shortcomings in this regard have, however, been more than made good by the magnificent series of photographs supplied by Mr. J. H. Court, the bulk of them from his own collection. Arranged, as they are, in chronological sequence, they provide in themselves informative and eloquent testimony of the advances made during the entire course of steam locomotive supremacy.

These photographs constitute an integral part of the present work. In them the discerning reader will be able to ponder at leisure the classical rigidity of the products of Crewe, the suave outlines of J. G. Robinson's creations for the Great Central, the elegance of S.W. Johnson's earlier work for the Midland and the perfect proportions of his later designs, the functional outlines of Churchward's locomotives (somewhat tamed in later years), the plain unobtrusiveness of the products of Doncaster and the glitter of a Wainwright 4-4-0 for the South Eastern and Chatham—to name only a few of the rich variety of subjects displaying British engineers' matchless ability to turn out a machine as pleasing to the eye as it was efficient in operation. (The author's sketches, drawn to scale but unencumbered with detail that would detract from their main outlines, serve the same purpose, particularly as regards nineteenth century work.) Finally, grateful acknowledgments are made to British Rail for supplying photographs of the latest developments of locomotive practice in this country. These productions combine the smooth lines characteristic of British methods with functional requirements in a manner that can hardly fail to commend itself, if not without some qualification, to the unprejudiced observer.

That the steam locomotive of the Stephenson pattern remained for so long the source of motive power on the railways—and, indeed, made railways a practical proposition—is a tribute to the soundness of the original conception. The extraordinary attraction it has exercised is not, perhaps, easily explained; but the fact is incontrovertible, and today, in its decline, it remains for many the most fascinating of all machines. It is to those who find it so, whether as onlooker or as having been professionally engaged in its design, construction, maintenance or operation, that the present volume is addressed.

G.G.

CONTENTS

ILLUSTRATIONS

Illustrations

CHAPTER I

Beginnings

THERE are many ways in which the history of the steam locomotive could be narrated. The developments which the basic Stephenson plan underwent could be recorded, for example, chronologically or with reference to each railway or each designer. But perhaps none of these would justify the re-telling of a familiar story. The advances, it is thought, can best be appreciated against the background of basic structure and plan to which the various developments conformed. This point of view—fundamental as it is to the understanding of the problems with which locomotive engineers have been confronted—has been curiously neglected, and is seldom adverted to in current descriptions, which in other respects are reasonably informative.

The period covered by the present chapter opens with the first engine to be built for the first public railway; before it closed the locomotive had assumed a form which was to remain fundamentally unchanged throughout the whole of its history. Crudities have been eliminated; but a coltlike awkwardness remains about even the most refined work of this opening period. Maturity has not been reached, but it is not far off. Nevertheless the story is not wholly one of uninterrupted development, as the above observations might suggest. The earliest years furnished several examples of development that were not pursued or did not survive the whole course. These too will be duly chronicled.

We propose first to describe *Locomotion* and its various developments; next the innovations which led to the production of the Stephenson engine, exhibiting all the characteristics of the modern locomotive, through in rudimentary form, then to describe the various developments which did not survive, and, finally, to give examples of the most advanced practice which the years preceding 1860 afford.

Early Efforts

Locomotion No. 1, built by R. Stephenson and Co. for the Stockton & Darlington, the first public railway to use steam power, must appear to those familiar with later developments as an odd contraption. The boiler, 10 ft. long and 4 ft. in diameter, had a single flue, 2 ft. in diameter, furnishing 60 sq. ft. heating surface, and was pressed to 50 lb. per sq. in. The two cylinders, 9 ½in. × 24 in., were sunk within it, each being in line with the two axles, which were coupled and actuated by the vertical pistons connected therewith by beam mechanism secured by upright supports fixed to the boiler. The cranks were set at 90°. Tribute should be paid to the ingenuity displayed in this pioneer work, but only a very short time was to elapse before its basic plan was to be improved out of all recognition.

Locomotion *1825* (*Simplified outline*)

Timothy Hackworth's *Royal George* of 1827 embodied the same general principles in a more developed form. The cylinders remained vertical but they were placed at one end of the engine, facing downwards and driving a dummy crank shaft connected by coupling rods to the six coupled wheels. The boiler, 13 ft. long and 4 ft. 4 in. in diameter, contained a double flue with a return bend, the chimney and firehole being side by side at the other end of the engine. The flue provided 141 sq. ft. heating surface and the boiler was pressed to 52 lb. per sq. in. Engines of this description required two tenders, one for coal at the leading end and the other for water at the rear.

A further stage of development was reached in Bolton and Leigh's *Lancashire Witch* of 1828—although, like *Locomotion*, this machine had only four wheels—in that the cylinders were no longer vertical but placed at an angle over the rear coupled wheels and arranged to drive the leading axle.

Hackworth's 0-6-0 of 1830 for the Stockton & Darlington, while resembling his *Royal George*, showed a real advance in boiler design in that the fuel fed into the central flue returned to the smokebox end through a series of tubes, flue and tubes communicating with a combustion chamber at the other end of the boiler. In their latest form Hackworth's engines incorporated inclined cylinders attached to the boiler, after the manner of *Lancashire Witch*.

A further advance in boiler design was made in Stephenson's famous *Rocket* of 1829. Here the internal flue was replaced by a firebox at the rear end of the boiler, covered at the top, back and sides with a water jacket and pierced in front by a series of holes into which a number of tubes communicating with the smokebox at the other end of the boiler were fitted. It only required the contour of the firebox casing to follow that of the barrel for the form of boiler with which the steam locomotive was associated for the rest of its career to be realized—and this was done in the later examples of the *Rocket* class. Unlike the engines previously described, which were employed on colliery work, the *Rocket* was a passenger engine, and coupling rods were not employed. The leading axle was the driver, the cylinders occupying a position over the carrying axle at the rear.

The *Rocket* had 4 ft. 8½ in. driving wheels and 8 in. × 17 in. cylinders. The barrel, 6 ft. long and 3 ft. 4 in. in diameter, contained 25 3-in. tubes. The firebox was 2 ft. long, 3 ft. wide and 3 ft. deep, with water spaces of 2½ in. at top, sides and back. The tubes provided 118 sq. ft. and the firebox 20 sq. ft. heating surface. The boiler pressure was 50 lb. per sq. in. and the total weight 4¼ tons.

In later engines of the series the diameter of the tubes was reduced to 2 in. and their number raised to 88 and then to 92. In the *Northumbrian*, the first to have the firebox and barrel following the same contour, the tubes were $1\frac{5}{8}$ in. in diameter and numbered 132. By this time the weight had risen to nearly $7\frac{1}{2}$ tons.

In these engines we find boilers of the modern type; but the position of the cylinders remains archaic. This was rectified in Stephenson's *Planet* of 1830 and in the well-known Bury type of the same year of origin, in which the cylinders were placed horizontally beneath the smokebox.

Developments

The question of the best position for the frames remained unresolved for many years. Ultimately single inside frames became universal but during our present period, the outside position was much favoured, combined with inside frames for part or the whole of the engine. Some designers, however, favoured outside frames for the carrying wheels and inside frames for the drivers. The use of plates was all but universal, Bury providing the only notable exception, in building up the frames from a series of bars.

The Bury Plan

Bury's engines were diminutive machines, even for these early days and were usually supported by four wheels only, coupled for goods work. The inside cylinders under the smokebox drove the crank axle of the second pair of wheels. The outer firebox, with hemispherical top and D-shaped on plan, lay behind the second axle. Edward Bury, who became locomotive superintendent of the London & Birmingham in 1837, had 88 of these at work by 1841, and it was not unusual for four of them to be used on the heaviest trains.

The 'Crewe' Frame

A pattern of framing much favoured in these early days is that associated with the names of William Buddicom and his assistant Alexander Allan of Crewe. The outside portion embraced the outside cylinders and supported the carrying, but not the driving, wheels, which had inside bearings only, attached to internal frames extending from the buffer beam to the firebox. The cylinders were thus firmly held by both sets of frames. Moreover, being outside, they eliminated the crank axle, always considered a source of weakness.

A typical engine of this pattern and the first built at Crewe, in 1845, of the 2-2-2 type had 18 in. × 20 in. cylinders, 6 ft. driving wheels, a total heating surface of 709 sq. ft., of which the firebox, with a grate area of 10·5 sq. ft., supplied 51 sq. ft. About one half of the total weight, 18 tons, rested on the driving axle.

Three 2–4–0 well tanks, supplied by Jones and Potts to the Shrewsbury & Hereford Railway in 1856–7, afford another example of this form of engine. These had raised fireboxes with Salter safety valves above them and at the centre of the barrel. They had no cabs and brakes were provided on the rear coupled wheels only. The boilers were fed by cross-head pumps. The chimneys had flared tops, and the square base, reminiscent of London & North Western practice. 15 in. × 20 in. cylinders drove the first pair of 5 ft. 6 in. coupled wheels.

William Barclay, a nephew of Alexander Allan, perpetuated this design of frame in the first two locomotives of the Inverness & Nairn Railway, subsequently incorporated in the Highland, upon which it remained standard practice for many years. These engines were of the 2–2–2 type and had raised fireboxes and safety valves placed over them and on the barrel as in the Shrewsbury & Hereford locomotives just described. They were fitted with four wheeled tenders, carrying 1,100 gals. of water and 2½ tons of coal. No brakes were provided on the engines, nor cabs, and the splashers, after the fashion of those days, were perforated. The driving wheels were 6 ft. in diameter and the cylinders 15 in. × 20 in.

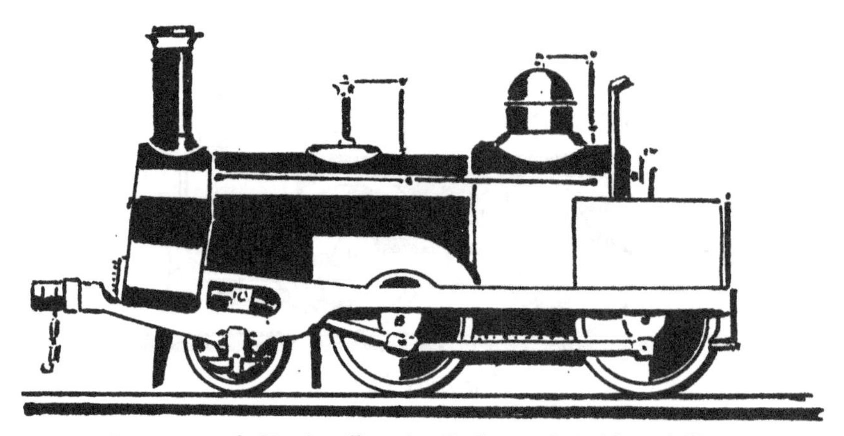

Inverness and Aberdeen Junction Railway 1858. 'Crewe' Frame

Stroudley rebuilt five of them as 2–4–0s with the larger, domed, boilers 3 ft. 7 in. in diameter, containing 181 1¾-in. tubes. The fireboxes, 4 ft. 6 in. long, provided 70 sq. ft. heating surface and 13½ sq. ft. grate area.

In 1858–9 Hawthorn's supplied the Highland with some 2–4–0 goods engines of the same general design, having 16 in. × 22 in. cylinders and 5 ft. wheels. Boilers, 4 ft. 1 in. in diameter, contained 218 1¾-in. tubes, and the fireboxes, 4 ft. 2½ in. long, provided 73¾ sq. ft. heating surface and 12¼ sq. ft. grate area. 9½ of the 28 tons total weight was borne by the leading axle. Two of these seven engines were subsequently provided with bogies at the leading end by Jones, Stroudley's successor.

The foregoing are far from exhausting the examples of the once popular Crewe framed locomotive, but they must suffice for the present purpose. Further examples are given in the chapter dealing with the mid-Victorian period. The Crewe frame provided a well knit locomotive but it was somewhat elaborate and, possibly with improved methods of manufacture, gave way to simpler forms.

1. Furness Tank Engine, 1866
2. Great Eastern 2–4–0 Johnson, 1868

3. Benjamin Connor's 2–4–0 for the Caledonian, 1872
4. Fletcher's North Eastern goods engine, 1875

Crampton Locomotives

We now proceed to consider another line of development which did not survive, and indeed had a much shorter life than the Crewe frame.

Early designers were convinced of the desirability of large driving wheels for express work and (with less justice) of the importance of keeping the centre of gravity as low as possible. The outcome of these convictions was the Crampton design, in which a single pair of large driving wheels was placed behind the firebox and preceded by two (in one case three) pairs of carrying wheels.

A weakness of this plan was the small proportion of the total weight available for adhesion. Thus in the locomotives built by Tulk and Ley for the Southern Division of the London & North Western in 1847 less than 12 tons of 25½ tons total weight rested on their 8 ft driving wheels. The outside cylinders measured 18 in. × 20 in. The boilers, pressed to 100 lb. per sq. in., were oval in section, 4 ft. 8 in. in the vertical and 3 ft. 8 in. in the horizontal plane. The grate area was 16 sq. ft. The frames were inside.

The Crampton System

Some Crampton engines supplied by Crewe for the Midland in 1848 had outside frames for the leading and driving wheels but inside frames for the intermediate wheels. 16 in. × 22 in. cylinders actuated 7 ft. driving wheels. The total heating surface was 1,062 sq. ft. and the grate area 13·9 sq. ft.

The position of the driving wheel behind the deep firebox in engines of this description prevented their being actuated by a crank axle, and outside cylinders were in consequence normally provided, but in some locomotives built by Longridge for the Great Northern, and delivered during Sturrock's reign at Doncaster (though not of his design), the Crampton plan was combined with inside cylinders. This entailed the employment of a dummy crank shaft, which received its motion through cranks from the inside cylinders and imparted it to the wheels by rods connecting the cranks at the outer ends with crank pins on the wheels.

Another, extraordinary, device which combined large wheels with low centre of gravity was adopted in the London & North Western 4–2–2 *Cornwall*, of 1847, in which the boiler was placed below the driving axle. This engine was subsequently rebuilt as a 2–2–2 with the boiler in the normal position, and ran for many years in this form.

The 'Long Boiler' Plan

Another unusual design of the period resembled the 'Cramptons' in that the driving wheels were placed behind two carrying axles but differed from it in that the firebox was placed behind the driving axle. This rendered engines of this description somewhat unsteady at speed and entailed the use of a long boiler—indeed this was a passenger version of the much favoured 'Long boiler' 0–6–0 goods engine, in which all the coupled axles were placed ahead of the firebox.

An example of this plan is furnished by some locomotives built by Jones and Potts for the Eastern Counties Railway in 1846. 6 ft. 8 in. separated the carrying axles and the wheelbase between the rear of these and the driving wheels was 5 ft. 4 in. in a total of only 12 ft. Over this was mounted a boiler 13 ft. 7 in. long and a 4 ft. 2 in. firebox. The former, 3 ft. 7 in. in diameter, contained 140 2-in. tubes, the latter, of gothic formation with the safety valves on top of it, furnished a grate area of 10½ sq. ft. The smokebox lay ahead of the first of the carrying axles, and the outside horizontal cylinders, 15 in. × 26 in., were between them. It is significant that these engines were subsequently rebuilt as 2–2–2s.

Six Wheeled Singles

This was unquestionably the most popular of single wheeler types during our period. Practice differed as regards the position of the cylinders and frames, which were in some cases outside and in others between the wheels.

One of the most famous engines of the period was *Jenny Lind*, built by E. B. Wilson in 1847 to the designs of David Joy. This engine, the prototype of many others, had outside frames for the carrying wheels and inside frames for the drivers. The cylinders were inside. The firebox was raised and a dome was placed on the centre of the barrel. Dimensionally the engine was not remarkable. 15 in. × 20 in. cylinders drove 6 ft. wheels. The total heating surface was 800 sq. ft., of which the firebox provided 80, and the boiler was pressed to 120 lb. per sq. in.

The 'Large Hawthorns' supplied by those builders to the Great Northern in 1852, may be taken as an example of the 2–2–2 with inside cylinders and outside frames throughout—a popular plan in those days. These engines had domeless boilers and raised fireboxes. 16 in. × 22 in. cylinders drove 6 ft. wheels. The boiler, 10 ft. × 4 ft., contained 171 2-in. tubes; the firebox, 5 ft. 1¼ in. long, provided 114 sq. ft. heating surface and a grate area of 13·6 sq. ft.; and they weighed 27 tons 16 cwt.

Contemporary engines of the same design built by R. Stephenson & Co. for the Midland had cylinders of the same size, 6 ft. 8½ in. drivers, a total heating surface of 1,097 sq. ft. (about 110 sq. ft. more than the Great Northern engines) and weighed 28 tons.

McConnell's practice on the Southern Division of the London & North Western was in complete contrast. Frames (as well as cylinders) were inside throughout. His earliest engines of 1851 had 16 in. × 22 in. cylinders, 7 ft. driving wheels, 195 2⅛-in. tubes, fireboxes (with transverse water partitions) furnishing 146 sq. ft. heating surface, and weighed 29½ tons. These excellent machines were followed by a smaller series a few years later and by others, to be described in the next chapter, of larger dimensions—McConnell's masterpiece—beyond our period. For some reason all these engines were called 'Bloomers'—'Large,' 'Small' and 'Extra Large' respectively.

The combination of inside frames and outside cylinders is exemplified by some engines built by E. B. Wilson's for the Eastern Counties Railway in 1847. These engines had gothic fireboxes. The cylinders were placed ahead of the leading wheels with the smokebox in line.

This necessitated a somewhat lengthy overhang at the leading end which combined with the outside cylinders tended to produce a lateral boxing movement at speed. They were small engines, with $15\frac{1}{2}$ in. × 20 in. cylinders, 5 ft. 6 in. driving wheels, a boiler of oval section, measuring 3 ft. 7 in. vertically and 3 ft. 4 in. horizontally and containing 113 2-in. tubes. The firebox, 4 ft. $0\frac{1}{4}$ in. long, provided 62 sq. ft. heating surface and a grate area of 9·8 sq. ft. They were subsequently converted into 2–4–0 saddle tanks for local work—a sphere of action more suited to their general proportions

A more famous class of this description was introduced by Ramsbottom on the Northern Division of the London & North Western in 1859, 60 of these being built in the ensuing 6 years. These engines, variously known as the 'Problem' or 'Lady of the Lake' class, were much in evidence on this railway for many years, latterly as pilots for the Webb three-cylinder compounds, which were not conspicuously speedy, as the engines being described, with their 7 ft. $7\frac{1}{2}$ in. driving wheels, undoubtedly were. They were small engines even for their period. The cylinders were 16 in. × 24 in. and $11\frac{1}{2}$ tons of their total weight (27 tons) rested on the coupled axle. 192 $1\frac{7}{8}$-in. tubes were provided, and the fireboxes furnished 85 sq. ft. heating surface and a grate area of 14·9 sq. ft. The boiler pressure was 120 lb. per sq. in. The fact that these engines were rebuilt towards the end of the century by F. W. Webb is a tribute to the soundness of their design and construction, though the uses to which they were frequently put, as above described, may be regarded as a reflection of their rebuilder's own work.

Sturrock's Giant

The most powerful single wheeler of our period was unquestionably that designed by Archibald Sturrock for the Great Northern and built by Hawthorn's in 1853. This engine, No. 215, had outside frames throughout and inside cylinders, 17 in. × 24 in., driving 7 ft. 6 in. wheels. The most notable feature was the size of the boiler, which was 4 ft. 4 in. in diameter and housed 240 2-in. tubes. The firebox casing, of the raised pattern, was 5 ft. 5 in. long (inside) and furnished 155 of the total heating surface of 1,719 sq. ft. The total weight was $37\frac{1}{2}$ tons. The absence of a dome anticipated Stirling practice, and the great size of its boiler for those days exemplified the principle subsequently enunciated by Ivatt of the same company that 'the power of an engine lay in its capacity to boil water', and applied by him in his large 'Atlantic' in 1902 and by Gresley, his successor, in his 'Pacifics' in 1922. The Great Northern tradition favouring large boilers was thus of long standing. Sturrock was also a pioneer in the adoption of high working pressure and it is believed that the boiler of this engine was pressed to 150 lb. per sq. in.

No. 215 was before her time. The work demanded of her was satisfactorily performed by his smaller 2–2–2s of the same general design. Sturrock asserted that his engine could readily have hauled trains between King's Cross and Edinburgh in eight hours. There is no reason to doubt the truth of this claim and it is unfortunate that the time table demands were not stepped up to take advantage of the power available. Too often the pace has been set by the traffic rather than the locomotive department and the potentialities of the railway, which, after all, owed its inception and development to the steam locomotive, have not been fully exploited.

2–4–0s

The 2–4–0, which was to attain great popularity in the mid-Victorian era, was well on the way to development to its final form in these early days. Normally the firebox was placed between the coupled axles but our period provides a number of examples of the long boiler variety, in which the firebox was behind the rear axle. The resulting overhang mitigated against the employment of these engines on any but slow traffic, while the length of the firebox was necessarily restricted.

A locomotive built by R. Stephenson & Co. for the Eastern Counties Railway in 1846 was of this description. Horizontal outside cylinders (15 in. × 22 in.), placed ahead of the leading wheels, balanced the overhang of the trailing end, entailed by the 4 ft. 2¾ in. firebox. The boiler, 13 ft. 7 in. long and 3 ft. 3 in. in diameter, contained 123 2-in. tubes. The total wheelbase was only 11 ft. 9 in. Inside frames were employed throughout and the firebox was of gothic formation. 6 ft. coupled wheels were used and the engine weighed just over 23½ tons.

Long-boiler 2–4–0 from the Eastern Counties R4 1848

E. B. Wilson's engines of 1853 for the Oxford, Worcester & Wolverhampton Railway, and the series designed by Sturrock for the Great Northern and built by Hawthorn's in 1855, may serve as examples of 2–4–0s with inside cylinders and outside frames—a very popular variety in the '50s and for many years later. Both had raised fireboxes. The cylinders of the former were 16 in. × 20 in. and drove 5 ft. 8 in. wheels. The boilers contained 184 2-in. tubes and the fireboxes provided 94½ sq. ft. heating surface and 15 sq. ft. grate area. The total weight was 32 tons—one ton more than the Great Northern engines, which had 16½ in. × 22 in. cylinders, 6 ft. 6 in. wheels, 4 ft. diameter boilers containing 160 2-in. tubes and a firebox heating surface and grate area of 110 and 14·9 sq. ft. respectively.

J. I. Cudworth's engines for the South Eastern, also of this general form with inside cylinders and outside frames, were noteworthy in that the firebox extended over the trailing coupled axle—an arrangement which greatly enlarged its length, though restricting its depth. In these engines the fireboxes were no less than 7 ft. 6 in. long and furnished 22·4 sq. ft. grate area and (with midfeather) 141 sq. ft. heating surface. In subsequent engines of the same class the fireboxes were shortened but the corresponding figures, 19·8 and 114, were large for the period. With 16 in. × 24 in. cylinders and 6 ft. coupled wheels, 120 lb. per sq. in. boiler pressure and a total weight of 30 tons, these engines compare favourably with their contemporaries on other lines.

Robert Sinclair's goods engines of 1859 for the Eastern Counties Railway may be cited as one example of the exactly opposite practice in that the cylinders were outside and the frames inside throughout. The dome, surmounted by Salter safety valves, was over the firebox which was not raised. 6 ft. 1 in. coupled wheels were actuated by 18 in. × 24 in. cylinders placed horizontally ahead of the leading wheels. The boiler, 4 ft. in diameter and pressed to 120 lb. per sq. in., contained 204 $1\frac{3}{8}$-in. tubes, and the firebox provided 65 sq. ft. heating surface and a grate area of 13 sq. ft. Altogether 110 engines of this class were built. In the later engines the domes were placed over the barrel and the dimensions varied.

Our period also witnessed the introduction of the well-known Beattie 2–4–0 on the London & South Western; but these belong substantially to the mid-Victorian era and are described in the next chapter.

The combination of inside frames and inside cylinders which, in the course of time, was to prevail as standard practice for this type of locomotive was not much favoured at this early stage, but some engines built by R. Stephenson & Company for the Chester & Birkenhead Railway in 1853 furnish an example. With $14\frac{1}{2}$ in. × 20 in. cylinders, 5 ft. 6 in. coupled wheels and 780 sq. ft. heating surface, they were small engines; but of more modern form than others considered above. A large dome was placed on the centre of the barrel and the firebox was raised.

0–4–2s

The front coupled six-wheeled locomotive won for itself considerable favour during the early years of locomotive history and continued to be employed on the Caledonian for goods work well into the mid-Victorian era. Important developments also took place with this type of locomotive for passenger and express work. The earlier examples with which we are now concerned were suitable only for slow moving traffic.

The familiar combination of inside cylinders and outside frames in locomotives of this period is exemplified in some 0–4–2s built by Hawthorn's for the Great Northern in 1848. These engines had raised fireboxes and domes on the middle of the barrel. They had no cabs or even weather-boards. The springs of the coupled wheels were compensated by a connecting beam. $15\frac{1}{2}$ in. × 24 in. cylinders drove the second pair of 5 ft. coupled wheels. The boiler barrel, 10 ft. long and 3 ft. 10 in. in diameter, contained 166 2-in. tubes and the firebox provided 75 of the total heating surface of 970 sq. ft. The total weight was 26 tons.

Patrick Stirling provides us with examples of two other forms of this type of locomotive. His first design for the Glasgow & South Western had outside cylinders and inside frames; in his second design for the same company both frames and cylinders were inside and the engines bore a strong resemblance to his later work for the Great Northern, besides anticipating the lines on which the 0–4–2 was to be developed at a later period on the London, Brighton & South Coast and London & South Western Railways.

Stirling's first engine for the Glasgow & South Western—supplied by Hawthorn's in 1856—had horizontal cylinders (16 in. × 22 in.) placed ahead of the (5 ft.) coupled wheels, with the smokeboxes in line with them. The domed boilers were 4 ft. in diameter and 11 ft. 2 in. long. Engines of similar design supplied by Neilson's in the following year had 142 2-in. tubes and fireboxes providing 53 of a total heating surface of 1,026 sq. ft.

Stirling's second design, with their inside cylinders, domeless boilers and safety valves housed in casings over the firebox were a far more characteristic product. 16 in. × 22 in. cylinders drove 4 ft. wheels. The total heating surface of 1,030 sq. ft. was furnished by 196 tubes and a firebox providing 76 sq. ft. having a grate area of $13\frac{1}{4}$ sq. ft. The boiler was pressed to 120 lb. per sq. in. and the total weight was 27 tons of which all but 22 tons were available for adhesion.

0–6–0s

This early period furnishes many examples of the variety of ways in which these axles all coupled can be combined with the Stephenson boiler and horizontal (or virtually horizontal) cylinders. The firebox may be situated behind the rear coupled axle or be placed between the last two. A third position, rarely adopted, is over the third axle. Cylinders may be inside or outside; so may the frames. Examples of representative types (to which in the interests of space our treatment must be confined) follow.

In this early period the 'long-boiler' engine was much in favour on colliery lines, since the short wheelbase facilitated the negotiation of sharp curves while the unsteadiness in running, due to the extensive overhang at the cab end, was of little or no disadvantage at the low speeds required. Moreover, the lengthy barrel made for economy in engines which spent a substantial part of their time standing.

Kitson's engines of this description for the Eastern Counties Railway, built in 1846, exhibited dimensionally the characteristics of the type. Boilers 13 ft. long were mounted on 4 ft. 9 in. wheels, spaced 6 ft. 3 in. and 5 ft. apart. Fireboxes overhanging the rear axle were 3 ft. 9 in. long. Cylinders were 15 in. × 24 in. and frames were inside.

William Bouch's engines of 1847 for the Stockton & Darlington were unusual in having outside cylinders—a feature which combined with the very short wheelbase, of 8 ft. 8 in., accentuated the inherent unsteadiness of the type. Another curious feature of the design was that the middle portion of the connecting rods was circular in form, in order to clear the coupling rod pins of the leading wheels—the arrangement being necessitated by the fact that the former were inside the latter. The cylinders were 16 in. × 24 in. and the wheel diameter was 4 ft. The boiler, 13 ft. 10 in. × 4 ft., contained 107 2-in. tubes, and the 3 ft. 5 in. firebox provided 10 sq. ft. grate area. The boiler was pressed to 80 lb. per sq. in., and the total weight was just over 25½ tons.

Still less common on this type of engine was the combination of inside cylinders with outside frames. The Stockton & Darlington possessed a single example, No. 86 *Zetland*, purchased in 1854, but built several years earlier by Hawthorn's. As usual the distance between the leading and middle axle was greater than that between the latter and the third—6 ft. 9 in. and 5 ft. 6 in. As rebuilt some 15 years after its acquisition, the engine had a boiler 12 ft. 10 in. long and 4 ft. in diameter. 170 1⅞-in. tubes were provided. The firebox was 4 ft. 2 in. long and the boiler pressure 120 lb. per sq. in.

The standard long boiler engines of the Stockton & Darlington and North Eastern Railways, which will be described in due course, had inside cylinders and frames.

Inside cylinders and frames were also generally favoured for the normal pattern of 0–6–0 with firebox lying between the second and third axles, though outside frames were common during this early period. Outside cylinders were exceptional.

Of the last, Patrick Stirling engines of 1855, built by Hawthorn's for the Glasgow & South Western, furnish our example. The cylinders, set horizontally, 16 in. × 21 in., drove 4 ft. 6 in. wheels. 48 of the 770 sq. ft. heating surface were furnished by the firebox and the grate area was 9½ sq. ft. These engines had domes and in practically every respect present a remarkable contrast to the same designer's later practice, which avoided the use of outside cylinders unless other features of design enforced their employment—which was certainly not the case where the 0–6–0 is concerned.

Sturrock's robust engines for the Great Northern provide a good example of the 0–6–0 with inside cylinders and outside frames. These were of the sandwich type with an ash centre bounded by iron plates. 16 in. × 24 in. cylinders drove 5 ft. wheels. The boiler was oval in section measuring 4 ft. 3 in. in the vertical and 4 ft. 1 in. in the horizontal plane. 187 2-in. tubes were provided and the raised firebox, 4 ft. 2 in. long, had a grate area of 14½ sq. ft. These engines, built by E. B. Wilson & Company in 1851–52, weighed 29½ tons. Others, generally similar, supplied by William Fairbairn & Son a few years later, had boilers of circular form, 4 ft. 4 in. in diameter, fireboxes a few inches longer, supplying 116 sq. ft. heating surface, and weighed 21 cwt. more.

Outside frames were also provided in R. Peacock's Manchester, Sheffield & Lincolnshire engines of 1852, which had the unusual feature of a firebox placed centrally over the third axle. This necessitated the incorporation of a bridge piece in order to secure the normal depth—an expedient which is said to have caused trouble in practice. They were very powerful locomotives for their time, with 18 in. × 24 in. cylinders and 5 ft. wheels. The 4 ft. 4 in. boilers, pressed to 120 lbs. per sq. in., contained 220 2-in. tubes and the firebox provided no less than 145 sq. ft. heating surface. The total weight was 33½ tons.

McConnell's Wolverton *Goods L.N.W.R. 1854*

Modern Form Achieved

An early example of the inside cylinder, inside framed 0–6–0, which was to become one of the most extensively used patterns of locomotive in this country is provided by McConnell's engines for the Southern Division of the London & North Western, which preceded Ramsbottom's better-known DX goods for the same railway by a few years.

McConnell's engines had unusually large wheels, 5 ft. 6 in. in diameter, but otherwise conformed to type. Introduced in 1854, more than 100 were built in less than ten years. The cylinders were 16 in. × 24 in. and the unusually high pressure for those days of 150 lbs. per sq. in. was adopted. The boiler, 4 ft. 4 in. in diameter, housed 234 1¾-in. tubes. The firebox, with a grate area of 16·3 sq. ft., contributed 109 of a total heating surface of 1,309 sq. ft., and the total weight was 26 tons 12 cwt.

The DXs, of which more than 900 were built at Crewe from 1858, had 17 in. × 24 in. cylinders, 5 ft. 2 in. coupled wheels, 1,102 sq. ft. heating surface, 15 sq. ft. grate area, boilers pressed to 120 lb. per sq. in., and weighed 27 tons.

With these engines we reach a stage of development that was to remain basically unchanged for many years and was, indeed, never to be wholly superseded.

CHAPTER II

Mid-Victorian

THE problems which confronted locomotive engineers during the period of thirty years beginning in 1860 were somewhat different from those which exercised their successors in late Victorian and Edwardian times. The limiting factor was not the maximum height and width permitted by the loading gauge, but the light weight capable of being carried by the permanent way of the period. The accommodation over the wheels of the Stephenson boiler, requiring as it did a deep firebox at the rear to ensure proper combustion, presented no difficulty. The diameter of the boiler was such that the transverse measurement of the inner firebox at the centre line of the barrel was invariably less than the width of the outer firebox at grate level. Hence the inner firebox could be inserted from below—a constructional convenience which had eventually to be abandoned. Moreover, these boilers could be readily accommodated in the space available between the tyres and enlargement of the driving wheels did not involve pitching the boiler unduly high (according to the notion of those days).

Much of the variation in later designs can be traced, as we shall see, to the manner in which locomotive engineers tackled these problems. Nevertheless, the period with which we are now concerned can show variations no less interesting. But these are due, not to the pressure of demands for greater power, but to the fact that we are still to some extent in the formative period of locomotive practice. The exploratory period is over, but the locomotive has not yet been moulded into a well-defined, universally accepted convention. For example, remarkable diversity of practice is evident in the position of frames and cylinders, no case more so than that of the six-wheeled 'single', which was responsible for much of the express haulage of the period and with which our survey may appropriately begin.

I THE SINGLE-WHEELER

These years furnish examples of 2–2–2s with inside and outside cylinders, and with single and double frames, the latter providing external support for one, two or all three axles.

2–2–2 *Outside Cylinders*

Sinclair's engines for the Great Eastern, Connor's for the Caledonian, and Sacré's for the Manchester, Sheffield & Lincolnshire had outside cylinders and double framing of the Crewe type, with outside bearings for the carrying wheels and inside bearings for the drivers.

The Caledonian engines, 16 of which were built between 1859 and 1875, were attractive machines. The centre-piece of the composition was the 8 ft. 2 in. driving wheels, fractionally, the largest ever used for a class on the standard gauge. Slits in the large splashers were reminiscent of the paddle-wheel steamers of the period. The outside horizontal cylinder was secured to external framing of the Crewe type, the cross-head, piston rod and big end being rendered visible, and accessible, by suitable openings in these frames. The smokebox was curved outwards to meet the cylinders, and a pleasing touch of austerity was provided by a stove-pipe chimney, which blended happily with the somewhat ornamental characteristics of the rest of the design. These engines had 17¼ in. × 24 in. cylinders, 192 1⅞-in. tubes, fireboxes providing 89 sq. ft. heating surface and 13·9 sq. ft. grate area, boilers pressed to 120 lb. per sq. in. and weighed just over 30½ tons, of which about 14½ rested on the driving wheels. They are reported to have worked very satisfactorily, though how such diminutive machines could have coped successfully with Beattock and other formidable gradients on the Caledonian lines remains a mystery.

Sinclair's Great Eastern engines, of which 31 were built between 1862 and 1867, were rather smaller, though they had the advantage of a larger grate. Sacré's 12 engines for the Manchester, Sheffield & Lincolnshire were considerably larger, turning the scale at 40½ tons, of which more than 17½ were available for adhesion. These engines had 17 in. × 26 in. cylinders, 7 ft. 6 in. driving wheels, grates furnishing 17 sq. ft. heating surface, and 150 lb. per sq. in. boiler pressure. They were the last of this general design and ended their existence on the Cheshire Lines service between Liverpool and Manchester.

Engines of this description were never very numerous. Nevertheless, they represented a definite stage in the development of the British locomotive, combining in their day the ancient and the modern and forming one of the most intriguing of the many patterns which, in the course of its long history, the steam locomotive assumed.

Even at this early stage it is doubtful whether British engineers would have favoured outside cylinders without some extraneous advantage derived from their employment, particularly where, as on six-wheeled engines with a comparatively short wheelbase, this imparted a 'boxing' motion at speed. This motive was to be found in the fact that the presence of a crank axle between the frames would have raised the boiler beyond a point then thought desirable.

2–2–2 *Inside Cylinders*

The majority of designers, however, preferred to employ inside cylinders with their six-wheeled singles. McConnell on the London & North Western (1861) used inside frames throughout, as did Stroudley on the London, Brighton & South Coast (1874). Outside frames with bearings for all axles were favoured by Sturrock on the Great Northern, Cudworth on the South Eastern, Kirtley on the Midland, and Armstrong and Dean on the Great Western. Stirling's Great Northern engines had outside bearings for the carrying wheels and inside for the drivers, while Holden of the Great Eastern used outside bearings in the leading wheels only—an arrangement derived from his 2–4–0 engines, which the 'singles' closely resembled.

McConnell's engines of 1861 were undoubtedly some of the finest and most powerful of their day. This was the final development, on similar lines, of earlier classes already described in the last chapter. McConnell was uninhibited as regards high-pitched boilers and the centres of his final 2–2–2s—the 'extra large' 'Bloomers'—were 7 ft. 5½ in. from the rail. The circular portion of the outer firebox was of larger diameter than the barrel, and a combustion chamber was provided which reduced the length of the tubes to 9 ft. 4 in. The firebox heating surface was 242 sq. ft., a phenomenal figure, and the grate area 25 sq. ft. About 14¼ of the 36¾ tons at which these engines turned the scale rested on the coupled wheels.

Stroudley's engines exhibited that designer's characteristics of neatness and excellent proportions and workmanship associated with all his creations. One, with 6 ft. 9 in. coupled wheels, built in 1874, was followed by 24 engines of similar design, but with 6 ft. 6 in. wheels, in 1880-82.

Of the engines combining inside cylinders and outside bearings throughout, the most powerful were those for which Dean of the Great Western was responsible. These had 18 in. × 24 in. cylinders, 7-ft. driving wheels, domeless boilers 4 ft. 2 in. in diameter, fireboxes furnishing 115 sq. ft. heating surface and 19·3 sq. ft. grate area, and weighed just over 36 tons, of which 16½ tons were carried by the driving wheels.

Stirling's Great Northern engines, which were developed in size and power during our period and a few years thereafter, were among the finest singles ever produced. In their intermediate form, during the '80s, they had 18½ in. × 26 in. cylinders, 7 ft. 7½ in. coupled wheels, 18·4 sq. ft. grate area and nearly 17 of a total weight of 39¾ tons available for adhesion. These may fairly be regarded as the most highly developed examples of the excellent *Jenny Lind*, designed by David Joy and built by E. B. Wilson in 1847.

Considerations of space preclude further mention of other notable engines in these groups, and attention must be directed to the developments occasioned by the substitution of a bogie for a single pair of leading wheels.

Bogie Singles

Of these the best-known were unquestionably Patrick Stirling's 'eight-footers', the first of which was the famous No. 1 constructed at Doncaster in 1870. These engines constituted a radical departure from that designer's normal practice which, in all his other designs (save that for a 0–4–4 tank), avoided the use of a bogie, and incorporated inside cylinders. The departure from these standards by the provision of outside cylinders for the reason above stated, was regarded as unavoidable with driving wheels of the maximum practical dimensions (8 ft. 1 in.), accompanied by a long (28-in.) piston stroke; and the employment of a bogie may well also be due to a desire to eliminate the lateral movement imparted to six-wheeled engines with cylinders in this position. However that may be, it is somewhat ironic that the engines for which Patrick Stirling is best known incorporated features which conflicted so remarkably with his usual practice.

These engines might be regarded as the direct descendants of some six-wheeled singles by the same designer, which appeared on the Glasgow & South Western a few years before this period opens.

So much has been written of the Great Northern engines that little can be usefully added here. But it may be noted that the design was developed during the mid-Victorian period, and thereafter, in successive batches. Dimensions varied, but those built about the middle of our period weighed just over 45 tons, of which 17 rested on the coupled wheels. The firebox, 6 ft. 2 in. long, provided 109 sq. ft. heating surface and the boiler was pressed to 160 lbs. per sq. in.

The contemporary 2–2–2s, previously noted, were designed to take the same boiler, though this was pitched with its centre 7 ft. 6 in. above the rail, 2½ in. higher than those of the bogie engines, smaller driving wheels notwithstanding.

In power these two classes were very similar. This is the verdict of a writer in the *Locomotive Magazine* in 1900 (Vol. V., p. 54): 'While cheaper both in first cost and in upkeep, these six-wheelers were found to be quite as efficient in the conduct of the express traffic. If anything they have proved themselves faster than the larger engines both as regards the maximum speed for individual miles and the average speed throughout a long run.'

The general plan of Stirling's 4–2–2s was adopted by Massey Bromley for a series of 20 engines on the Great Eastern constructed between 1879 and 1882. But the Stirling long stroke was not incorporated, 24 in. being favoured in combination with 7 ft. 6 in. wheels. The valves were placed over the cylinders. The firebox furnished 110 sq. ft. heating surface and 17·1 sq. ft. grate area, and the engines weighed just over 41½ tons, of which little more than 15 tons were available for adhesion. (Plate No. 8).

Some years later, in engines of the same wheel arrangement, the Great Eastern was to follow the practice of another of the great trunk lines for its single wheelers, in this case the Midland, with the result that the two classes—one with outside cylinders and inside frames throughout, the other with inside cylinders, double frames for the driving and outside bearings for the trailing wheels—bore little resemblance, save as regards wheel arrangement, to each other.

But these latter engines belong to the '90s and will be noted in due course. So also will their Midland prototypes, though the first eight of the 95 engines of which they were the fore-runners appeared in the late '80s. Our survey of this part of our subject may fittingly conclude with the famous Caledonian No. 123, still happily with us (Plate No. 9). Here we find a bogie single of the most authentic British tradition with inside cylinders and frames throughout, the only example of such a machine which this period furnishes. Cylinders 18 in. × 26 in. drove 7-ft. coupled wheels. The boiler was 10 ft. 3 in. long and 4 ft. 3 in. in diameter (mean). A total heating surface of 1,053 sq. ft. was considerable for the period. The firebox provided a grate area of 17·5 sq. ft. and 111 sq. ft. heating surface.

The engine, which was built by Neilson's in 1886, weighed 41¾ tons, of which 17 were available for adhesion. No. 123 was basically a Drummond conception, though it has been disputed how far, if at all, that locomotive engineer was responsible for her. Built, and it is said designed, by Neilsons in 1886 and purchased by the Caledonian, the general outline of this engine and the great similarities she presents to the contemporary Drummond machines on the same railway—to say nothing of his earlier work on the North British and his later work on the London & South Western—would suggest that a denial of Drummond inspiration is pedantic. Differences there may be, but on a broad view No. 123 must be pronounced an authentic member of the Drummond family.

The contrast presented between No. 123 and even some of the best productions of an earlier part of the mid-Victorian period illustrates the immense strides that had been made during a quarter of a century. Take, as an example of earlier practice, McConnell's 'singles', to which reference has already been made. These engines, certainly among the most advanced of their time, with weatherboards for cabs and all controls visible, ornate boiler mountings, and open splashers, would have shocked late Victorian, to say nothing of Edwardian, susceptibilities by their crudity. Yet No. 123, eighty years later, excites no comment, save for her diminutive size. She is the equal as regards lines of the latest products of British engineering skill. The McIntosh 4–6–0s of the early years of the twentieth century or the superheated 140s of the 1920s are but enlarged versions of the same general plan, and it is not extravagant to say that, before the end of the '80s, the British locomotive, as exemplified by No. 123, had reached a stage of development which may be regarded as its most characteristic British form.

II SIX WHEELS FOUR-COUPLED

The great majority of (tender) locomotives constructed in our period (1860–90) were carried on six wheels. We have already dealt with the uncoupled variety and those in which all the wheels were coupled are considered hereafter. In the present section we are concerned with locomotives of an intermediate type, in which four coupled wheels were associated with a single carrying axle. Engines within this description were used for every variety of traffic, from fast express to slow goods work. Only two variations are, of course, possible—the carrying axle must either precede or follow the drivers. It is proposed first to consider engines of the 2–4–0 type and then those of the 0–4–2 formation; but a preliminary word on the characteristics of each will be appropriate. In the former a deep firebox, ensuring adequate combustion, was accommodated between the coupled wheels and the cylinder was placed over, usually somewhat ahead of, the carrying axle, providing adequate room for the machinery and ensuring direct exhaust through the smokebox above them. In the 0–4–2, greater latitude for the firebox was obtained, since its length was not restricted by the practical length of the coupling rods; but in engines of the dimensions then regarded as feasible this was of minor importance, while the cylinders had to be slightly inclined to allow the machinery to clear the leading coupled axle, and the guiding influence of a carrying axle at the front of the locomotive had necessarily to be sacrificed. It is not surprising, therefore, that for fast work the majority of designers preferred the 2–4–0 type; and it will be convenient first to draw attention to some notable examples of this wheel formation.

2–4–0s

Locomotives of this type were employed on almost every railway of Great Britain and, in order to keep our treatment within reasonable bounds, it must be emphasized that attention is directed only to the principal railways and to the most notable classes on each of them.

The 'Premier Line', as the London & North Western, not without reason, claimed to be, was an extensive user of 2–4–0s, which formed, indeed, the backbone of its express locomotive stock in those days. These were of four classes, Ramsbottom's 'Newtons' and 'Samsons' and Webb's 'Precursors' and 'Precedents', which numbered respectively 96, 90, 40 and 70 engines. Of these four classes the last named, developed from the 'Newtons', were the most important. The 'Samsons' were a smaller edition of the 'Newtons' and the 'Precursors' resembled the 'Precedents' save for being provided with 5 ft. 6 in. coupled wheels, which rendered their employment on express duties short-lived. A further series of 'Precedents' was produced by replacements of the 'Newtons' with locomotives bearing their names and numbers, but little, if anything else of the originals. The final chapter took place after the close of our period by the production, between 1891 and 1895, of a new series of 'Precedents', which were officially replacements of the engines whose names and numbers they bore.

By no means large as regards dimension for their day—the first was built in 1874—the 'Precedents' were one of the 'great' engines of all time. Inside cylinders, 17 in. × 24 in., drove the leading pair of 6 ft. 7½ in. coupled wheels, which were placed with their axles 8 ft. 3 in. apart, allowing a firebox 5 ft. 5 in. long, furnishing 103 sq. ft. heating surface and 17·1 sq. ft. grate area, to be placed between them. A barrel of 4 ft. 2 in. maximum diameter housed 198 1⅞-in. tubes, and the engine weighed 32¾ tons, of which 22½ rested on the coupled wheels. The boiler was pressed to 140 lb. per sq. in. The steam chest in section was an inverted triangle with the valves working on the inclined sides, and much of their excellence was doubtless attributable to the ample steam chest volume and free exhaust obtained by this arrangement.

Charles Rous-Marten paid the following tribute to these engines in *The Railway Magazine* of January, 1903: 'Here we have the problem of a tiny pocket-pistol of an engine . . . costing, I have been told, only about £1,800, performing with complete efficiency duty that on many other lines has demanded locomotives costing nearly, if not quite, double as much, weighing half as much again, and with 50 per cent. more theoretical power, to perform with equal efficiency. Surely, in view of these facts, which are established indisputably, Mr. Webb's 'Precedents' must be deemed *relatively* the most successful engines British railways have ever seen. Yes, I really think so!'

Webb's later designs having this wheel arrangement (although not falling strictly within it, as the two pairs of driving wheels were uncoupled), incorporating as they did his three cylinder compound system, did not meet with the same success. The system involved the employment of one low-pressure cylinder placed below the smokebox and two outside high-pressure cylinders, coupled to the front and rear driving axles respectively. The first engine, appropriately named *Experiment*, appeared in 1878 and was joined in the '80s by 29 others; a second class, the 'Dreadnoughts', consisted of 30 engines, and a third, the 'Teutonics', of ten. The 'Experiments' had 'Precedent' boilers, and the leading axle and chimney were in line. The other two classes had longer boilers, with the smokeboxes and leading axles in the customary relative positions, and advantage was taken of the absence of coupling rods to place the driving axles 9 ft. 8 in. apart and fit a longer firebox furnishing 20·5 sq. ft. grate area.

The 'Experiments' were failures, and the 'Dreadnoughts' only a little better, but the 'Teutonics' proved to be good engines—incomparably the best passenger compound engines (three- or four-cylinder) for which Francis Webb was responsible. One of them, *Adriatic*, on the last night of the Race to Aberdeen in 1895, covered the 158 miles between Euston and Crewe at an average speed of 64·8 miles an hour—a marvellous performance notwithstanding the light load (75 tons); but even this was outclassed by the average of 67·1 miles an hour on the same night, over the harder road between Crewe and Carlisle, achieved by *Hardwick*, of the 'Precedent' class.

The neighbouring Midland, which obtained access to London over its own metals in 1868 and by its enterprise became a formidable competitor to the North Western, was also an extensive user of 2–4–0s. Its locomotive policy presented a remarkable contrast to that of the premier line and that of its eastern neighbour, the Great Northern, in that standardization was not pursued with the same rigidity. Each fresh batch of locomotives, even when of the same general design, had marked characteristics of its own; and this renders anything approaching a detailed study of the large group of engines of the 2–4–0 formation out of the question. As late as 1906, when the Midland locomotive stock was renumbered, 2–4–0s, which contributed Class I, mustered nearly 300 engines. They were of great variety, both in form and dimensions, and for our purpose it must suffice to give a short description of two examples as indicative of the general character of this large class of locomotives.

Matthew Kirtley, who was in charge of the Midland locomotive department during the first half of our period, designed some excellent engines. They were very well built and their length of life was prodigious, many of them outlasting the Midland itself. This longevity was in large measure due to their robust construction, to which their double frames undoubtedly contributed. They were, however, somewhat ungainly and built at a time before many of the railways had evolved external characteristic lines stamping them unmistakably as the product of the owning company.

S. W. Johnson, who succeeded Kirtley in 1875, besides being an excellent engineer, was a master of this art. None excelled, and few equalled, him in producing locomotives of pleasing outline, and a double-framed Kirtley with a Johnson boiler carrying the latter's inimitable mountings was something of a hybrid.

Not the least notable of these products was the former's '800' class, which we take as the first example of the numerous Midland classes of the 2–4–0 wheel formation.

Built in 1870–1 and fitted with Johnson boilers from 1875 onwards, these 48 engines had 18 in. × 24 in. (26 in. in a few cases) cylinders and 6 ft. 9 in. coupled wheels. The boiler, of 4 ft. 3 in. maximum diameter, contained 223 1¾-in. tubes, 10 ft. 6 in. long. Firebox heating surface and grate area were 110 and 1,705 sq. ft. respectively. The boiler was pressed to 140 lb. per sq. in. (subsequently increased to 160 in a few cases) and the total weight was 40½ tons.

The other class to be mentioned, built in 1879 and 1880, was a typical Johnson product, exemplifying all that designer's artistry. The leading wheels were supported by outside bearings, the main frames being inside. Cylinders 18 in. × 26 in. drove 6 ft. 8½ in. coupled wheels. The firebox furnished 110 of the total heating surface of 1,206 sq. ft. and 17·5 sq. ft. grate area, and the total weight was 38 tons.

Crossing the road, from St Pancras to King's Cross, we find in Patrick Stirling another extensive user of the 2–4–0. Engines of this wheel arrangement were the subject of his first design and several series were built during our period. After certain modifications, they followed a rigid pattern and formed one of this designer's standard classes. Unlike the Midland and North Western engines the Great Northern 2–4–0s were not normally employed on the fastest work, which was entrusted almost entirely to single wheelers. With their 6 ft. 7½ in. driving wheels (one foot less than their companion 2–2–2s) they may be regarded as passenger rather than express locomotives. The final series of these engines, the first of which appeared in 1888, had 17 in. × 26 in. cylinders, boilers 4 ft. 0½ in. in diameter outside the smallest ring, containing 174 1¾-in tubes, and fireboxes 5 ft. 6 in. long, providing 16¼ sq. ft. grate area and 92¼ sq. ft. heating surface. Boiler pressure was 160 lb. per sq. in. and 27½ of the 39 tons total weight were available for adhesion. Stirling built 139 2–4–0s, of which the final series consisted of 56 engines.

Great Northern territory ended, as it was said, 'in a ploughed field' a few miles south of York. Northward, Edward Fletcher reigned over North Eastern locomotive matters in benevolent autocracy. This railway—a far larger affair than its southern partner—inherited from its constituents locomotive shops at York, Leeds, Darlington and Gateshead (where Fletcher resided), and the origin or headquarters of North Eastern locomotives was readily ascertainable from certain characteristic features and the manner in which they were painted. Dark brown underframes denoted Darlington, claret York, and an all-over green Leeds. Even the dominant colour common to them all (green) varied in hue. These differences, moreover, extended to mechanical details and boiler mountings, and may be cited as examples of the latitude allowed to, perhaps, the most independent set of enginemen in Great Britain, as Fletcher's successor was to find to his cost.

Fletcher himself was the second locomotive chief of a large company (the first was Beattie of the London & South Western) to abandon the single-wheeler in favour of the coupled engine for express work, and his 2–4–0s were the mainstay of North Eastern express locomotive power during our period. They were of a great variety, but a series of more than 50 produced between 1872 and 1882 may be taken as illustrative of their ultimate development. Cylinders 17 in. (17½ in. in some of them) × 24 in. drove 7-ft. coupled wheels. The 254 1$\frac{9}{16}$-in. tubes were housed in 4 ft. 3 in. diameter boilers, pressed to 140 lb. per sq. in. Fireboxes provided 16·1 sq. ft. grate area and 985 sq. ft. heating surface, and just over 27 of a total weight of all but 39½ tons was available for adhesion. These engines, which had inside cylinders and main frames, with the leading axle supported by outside bearings were of robust construction. Fletcher used, for example, frames 1¼ in. thick while Webb was content with a thickness of ⅞ in.

Towards the end of this period, after the short McDonnell interlude, they were supplemented by the admirable 'Tennants', officially produced by the board and so called after the general manager, but probably in substance the work of T. W. Wordsell, who was to succeed as locomotive chief. These engines resembled Stirling's Great Northern creations in many respects, but were provided with domes, and earned for themselves a reputation for speed usually associated with the last-named designer's single-wheelers.

Cylinders 18 in. × 24 in. drove 7 ft. 1¼ in. coupled wheels. The 4 ft. 3 in. diameter boiler housed 205 1¾-in. tubes. Firebox heating surface and grate area were 107 and 17·3 sq. ft. respectively. The boiler was pressed to 160 lb. per sq. in. and the engines weighed 42 tons, of which 29½ rested on the coupled wheels.

Wordsell himself was subsequently to produce a two-cylinder compound 2–4–0, with inside bearings throughout, but it remained a solitary example, the remaining engines being provided with a bogie at the leading end—which was hardly surprising, having regard to the additional weight entailed by the provision of a large low-pressure cylinder.

The Great Northern's other neighbour, the Great Eastern, possessed some excellent 2–4–0 engines designed by James Holden—one for express traffic with 7-ft. coupled wheels, the other for general purposes with smaller, 5 ft. 8 in. diameter, wheels. Like the Great Northern, North Eastern and (later) Midland engines, they had outside frames for the leading and inside frames for the coupled wheels. Cylinders, 18 in. × 24 in., were inside. The boilers, which were identical for both series, in their latest form were built up of two rings with the dome placed near the chimney on the first of them. 4 ft. 4 in. in diameter, they contained 254 1⅝-in. tubes. The smokeboxes furnished 110 sq. ft. heating surface and 18 sq. ft. grate area. The express engines weighed 42 tons. More than 200 of these engines were built and formed the backbone of Great Eastern stock for many years. With their stove pipe chimneys, neat but adequate cabs and their effective finish in blue, set off by scarlet coupling rods, they presented a characteristic appearance which marked Great Eastern productions until the end of the century. They followed the tradition set by T. W. Wordsell in an earlier series (1882–3)—a development of S. W. Johnson's small 2–4–0 of 1867 (Plate No. 2)—and differed from Sinclair's productions of 1859–61, which had outside cylinders and were of an altogether more slender character (intended for goods, but largely used for passenger work).

Retracing our steps from Liverpool Street, past King's Cross, St Pancras and Euston, we arrive at Paddington, where, again, we find the 2–4–0 much in evidence, particularly as maids of all work.

The Great Western locomotive stock of this type resembled that of the Midland (and probably surpassed it) in the variety of forms it took, and here also we must content ourselves with a short description of a few typical examples, representative of the work of Joseph Armstrong, at Wolverhampton, and William Dean, at Swindon.

This railway was much addicted to the sandwich frame, built up of wood bound by metal plates. The leading wheels were usually supported by outside bearings. For the coupled wheels inside frames were used in some cases and outside frames in others. Apart from frame design the boilers reflected different policies pursued from time to time. Barrels were composed of two or three courses with domes—set centrally, near the chimney or well back—or without domes. Some had raised, others flush, fireboxes. Moreover, these boiler variations might be combined with different types of frame, and the result was legion. Indeed, a complete history of the Great Western 2–4–0s would require a volume in itself.

Our first example may be taken from Armstrong's 'reconstruction' of an earlier series—in fact virtually new engines, built at Wolverhampton in 1885. Provided with inside cylinders and frames but with outside bearings and springs for the leading wheels, these locomotives had 17 in. × 24 in. cylinders and 6 ft. 2 in. coupled wheels. Boilers pressed to 140 lb. per sq. in. contained 248 1⅝-in. tubes. Firebox heating surface and grate area were 99 and 15½ sq. ft. respectively, and the total weight was 44 tons.

The next example illustrates a combination of Wolverhampton and Swindon practice. These, the 'River' class, derived from names subsequently given to them, began life as 2–2–2s in the fifties. These were subsequently reconstructed, in 1872, at Wolverhampton, retaining frames of the sandwich type, and provided with 17 in. × 26 in. cylinders and 6 ft. 8 in. coupled wheels. When provided with Swindon boilers they showed some advantage over the engines above described. The barrels, of 4 ft. 3 in. maximum diameter, contained 245 1¾-in. tubes and the fireboxes furnished 101 sq. ft. heating surface and 17·66 sq. ft. grate area.

Dean's last series, which may be taken as our third example, were an altogether more modern product. Built in 1892–3 (beyond our period, but typical of it), they had inside frames with the usual outside support for the leading wheels. Boilers, pressed to 150 lb. per sq. in., were of the Swindon pattern just described. Inside cylinders 17½ in. × 24 in. drove 6 ft. 8½ in. wheels and they weighed nearly 40 tons, of which all but 12 were on the coupled wheels.

On the railways south of the Thames, where the work required of the locomotives was generally of a less exacting character than that on the northern lines, the most extensive user by far of 2–4–0s was the London & South Western. Here Joseph Beattie's slender creations held sway for many years. This designer was the first to abandon the single wheeler for the coupled engine and his 2–4–0s were the mainstay of South Western stock for passenger service until 1876, when 4–4–0s were introduced and began to supersede them.

In general design Beattie's engines present a remarkable contrast to most of those already described. Outside horizontal cylinders were combined with inside frames and bearings, for all the axles, though the leading wheels were provided with light auxiliary bearings secured by the spring supports to the guide bars. The same designer showed great ingenuity in boiler design and feed-water heating contrivances, the latter in the most recent engines entailing the use of a small diameter chimney placed immediately ahead of the main chimney, each provided with an ornamental brass or coloured cap. Indeed, these engines, with their perforated splashers, curved running plates, and other features, were the most ornamental machines in their day, which was far from being inclined to severity in locomotive outlines.

They were built in several batches of differing dimensions, including the coupled wheels; but the following, which apply to a series built at Nine Elms in 1868, may be cited as typical. Cylinders 17 in. × 22 in. actuated 7-ft. coupled wheels. The total heating surface was 1,102 sq. ft. Grate area was 18·8 sq. ft. and boiler pressure 130 lb. per sq. in. The total weight was only 35½ tons. These comparatively small engines—even for the period (the last series, of modified design, was built under the superintendency of W. G. Beattie in 1873)—seem to have met the not very exacting South Western locomotive requirements quite satisfactorily.

The South Eastern obtained from Sharp, Stewart & Company, in 1875, a series of 20 2–4–0s, modelled on the North Western 'Precedents', which augmented Cudworth's creations on that railway. 17 in. × 24 in. cylinders drove 6 ft. 6 in. wheels and, with boilers 4 ft. 1 in. in diameter, pressed to 140 lb. per sq. in., and fireboxes 5 ft. 1 in. long, they weighed only 33 tons, being known as 'Ironclads' notwithstanding (Plate No. 5).

The London, Chatham & Dover was a moderate user of 2–4–0s, built between 1861 and 1873, for which it favoured outside frames and inside cylinders. The last series had 17 in. ×

5. Watkin's 2–4–0 for the South Eastern, 1875
6. Kirtley's tank engine for L.C. & D.R., 1875

7. Tank engine for the North British by D. Drummond, 1879
8. Great Eastern single by Massey Bromley, 1879

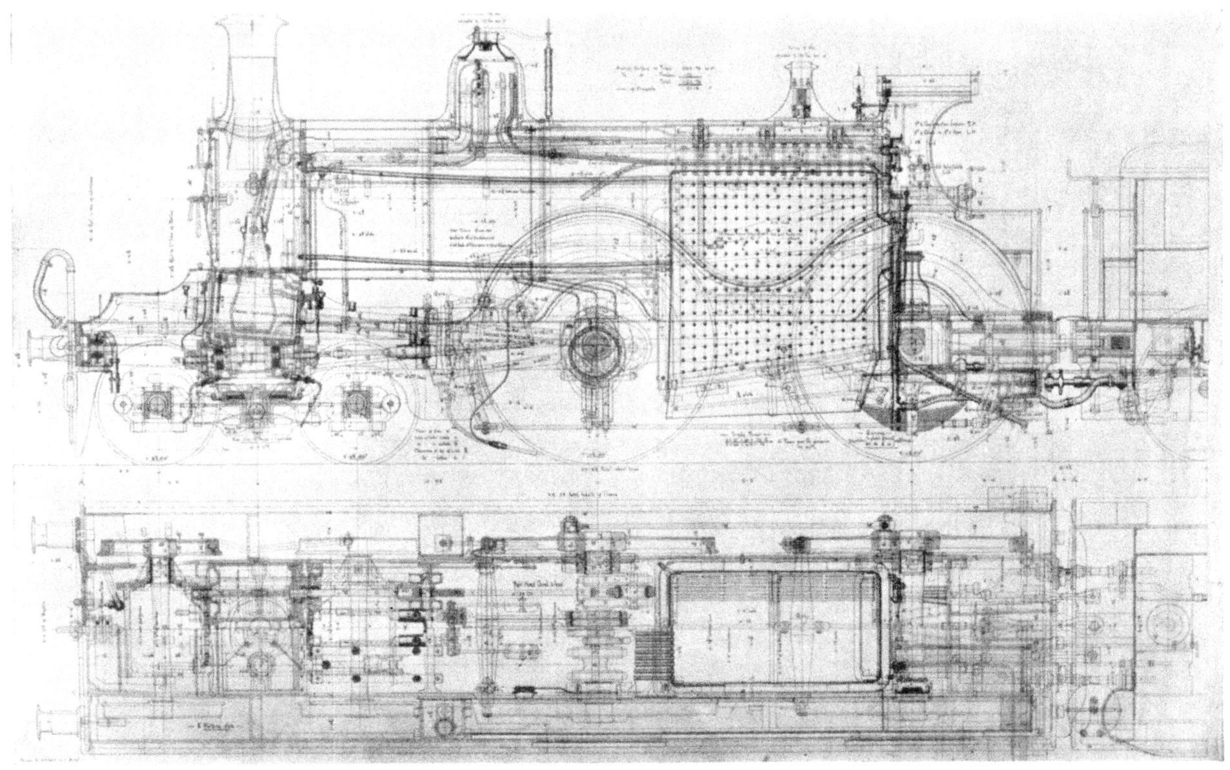

9. D. Drummond's Neilson-built engine for the Caledonian 1886
10. Johnson 4–4–0 for the Midland. Elevation and plan

11 & 12. A study in grace: typical Johnson creations for the Midland

24 in. cylinders, 6 ft. 6 in. coupled wheels, 1,121 sq. ft. total heating surface, 160 lb. per sq. in working pressure, and weighed 36½ tons.

On both these railways 2–4–0s were replaced by bogie engines at a relatively early period. On the London, Brighton & South Coast the 2–4–0 wheel arrangement was abandoned in favour of the 0–4–2 by the famous William Stroudley, though the first two of that designer's express engines, *Belgravia* and *Goodwood*, were of the 2–4–0 order. These engines had 17 in. × 24 in. inside cylinders and 6 ft. 6 in. driving wheels. Boilers of 4 ft. 3 in. diameter contained 260 1½-in. tubes. Fireboxes, 6 ft. 2 in. long, furnished 112 sq. ft. heating surface, 28¼ of the total weight of 41¼ tons being available for adhesion. They were transitional hybrids, boilers (domeless though they were when new), chimneys and cabs were genuine Stroudley, but below the running plate, with their outside frames, cranks and springs, they were reminiscent of Craven's miscellaneous creations.

Few would quarrel with the awarding of the palm for Scottish 2–4–0s to Hugh Smellie's 'Twelve Apostles', built at Kilmarnock in 1881 for the Glasgow & South Western. There were twelve of them, but the reason for the second part of the appellation is less obvious. Certainly—like all this designer's productions—they were exemplary locomotives—fast, powerful and economical. They had inside frames throughout. Cylinders, also inside, 18 in. × 26 in. drove 6 ft. 9½ in. coupled wheels. The boilers, 4 ft. 2 in. in diameter and pressed to 140 lb. per sq. in., contained 240 1⅝-in. tubes. Grate area and firebox heating surface were respectively 16 and 101 sq. ft., and the engines turned the scale at 38½ tons, of which 27 rested on the coupled wheels. In one sense these engines constituted a reversion to earlier practice, James Stirling having introduced bogie engines for the main line express work in 1868 as a development of his earlier engines of the 2–4–0 order.

The North British also employed the 2–4–0 wheel arrangement with inside cylinders and frames throughout under the aegis of Thomas Wheatley, who was the first to replace engines of this type by inside cylinder 4–4–0s of a design that was to provide the ground-plan for British express locomotives for many years. Connor on the Caledonian favoured outside cylinders and Crewe frames (Plate No. 3).

It should also be recorded that the 2–4–0 was a wheel arrangement much favoured by the minor railways, such as the Cambrian, Furness, North Staffordshire, and Midland & South Western Junction; but these engines were of a conventional pattern and cannot be further described in this section, which has already assumed very substantial proportions.

0–4–2s

The story of the 0–4–2 is more rapidly told. Its origins lie in a form of locomotive much used in Scotland for goods traffic and providing an example of the somewhat rare combination of outside cylinders with front coupled wheels—an arrangement which results in a somewhat heavy front end, rendering the locomotives unsuitable for any but slow-moving traffic. The more conventional form, with inside cylinders sited just ahead of the leading coupled axle, as opposed to the wheel, was much favoured during the mid-Victorian period for all kinds of work. Patrick Stirling used it for goods traffic on the Great Northern. His brother James, of the Glasgow & South Western, provided an excellent example for mixed traffic; so did Adams of the London & South Western, for general passenger work. But undoubtedly the greatest exponent of the art was William Stroudley of the London, Brighton & South Coast, whose 0–4–2s for mixed and express traffic achieved fame, to which their canary finish may have contributed but by no means fully accounted for.

A few typical examples will suffice to complete this part of the subject.

The outside cylinder goods engine was, as has been said, the product of an earlier age, but the Caledonian continued to produce locomotives of this description down to 1871. The following leading dimensions apply to the latest batch: cylinders 17 in. × 24 in., coupled wheels 5 ft. 2 in. diameter; heating surface—firebox 83 sq. ft., total 298 sq. ft., grate area 13·3 sq. ft. A lightly-loaded rear axle carried 4 tons 7 cwt. of a total weight of just over 30 tons.

Patrick Stirling's 0–4–2s formed one of that designer's standard types and exhibited all his characteristic features, including domeless boilers and extreme neatness, not to say plainness, of outline. With their 5 ft. 7 in. coupled wheels they constituted an intermediate class between the 6 ft. 7 in. 2–4–0s, already described, and the more powerful 0–6–0 goods engines reserved for notice hereafter. Boilers of the latest series, built from 1887, 4 ft. 0½ in. in diameter outside the smallest ring and 10 ft. long, contained 174 1¾-in. tubes. The firebox, 5 ft. 6 in. long, provided 92·4 sq. ft. heating surface and 16¼ sq. ft. grate area. Cylinder dimensions and boiler pressure were 17½ in. × 24 in. and 160 lb. per sq. in. respectively, and the engines weighed just over 35 tons, of which the trailing wheels carried about 8¼.

James Stirling's Glasgow & South Western engines bore a strong family likeness to his brother's Great Northern creations, points of resemblance including a severity of outline typified by the absence of a dome, and safety valves (in this case of the uncovered Ramsbottom type) over the firebox. The series built in 1864 had 16 in. × 22 in. cylinders, 5-ft. coupled wheels, 120 lb. per sq. in. working pressure and 13⅓ sq. ft. grate area. Boilers of 4 ft. 0¾ in. maximum diameter housed 162 2-in. tubes and the firebox provided 78 of a total heating surface of 930 sq. ft. A larger series of the following year had 17 in. × 24 in. cylinders and 5 ft. 1 in. coupled wheels.

Adams's 'Jubilees', on the South Western, which derived their name from 1887, their year of introduction, had 18 in. × 26 in. cylinders and 6-ft. coupled wheels. Boilers, pressed to 160 lb. per sq. in., 4 ft. 4 in. in diameter, contained 216 1¾-in. tubes. Fireboxes, 6 ft. long, furnished 110 sq. ft. heating surface and 17 sq. ft. grate area. They weighed 43 tons 5 cwt. These engines represent an important departure from their designer's practice of employing outside cylinders, an impractical position (as has been said) for engines of this type for other than the slowest work. 90 were built down to 1895.

Stroudley's 'Gladstones'—one of the great locomotive classes of all time—were preceded by a mixed traffic class introduced in 1876. Cylinders 17 in. × 24 in. actuated 5 ft. 6 in. coupled wheels. Boilers 4 ft. 3 in. in diameter (maximum) contained 202 1¾-in. tubes, for which 244 1½-in. tubes were substituted in the later engines. Fireboxes 5 ft. 8¼ in. long provided 103 sq. ft. heating surface and 17 sq. ft. grate area. Just over 27 of the 34 tons 7 cwt. total weight were available for adhesion.

An intermediate class, weighing 36 tons (28 tons adhesion), with 17½ in. by 26 in. cylinders, 6 ft. 6 in. coupled wheels and boilers of similar dimensions to the engines just described, first appeared in 1878.

'Gladstone', the first of the latest series of 36 engines—Stroudley's *magnum opus*—was built in 1882. The cylinder diameter was enlarged to 18¼ in. The boiler was 3 in. larger in diameter and the firebox 1 ft. longer. 333 1½-in. tubes were provided (Stroudley had no inhibition against a crowded barrel, of which the excellent standard of maintenance he insisted upon may furnish a partial explanation), and firebox heating surface and grate area were respectively 113 and 20·65 sq. ft. The total weight was 38 tons 14 cwt., of which 28 tons 6 cwt. rested on the coupled wheels.

III FOUR-COUPLED BOGIE

Locomotive observers of the mid-Victorian period with a forward look might well have regarded the genesis and development of the inside cylinder 4–4–0 for express traffic as its most significant feature. So it proved to be, for engines of this pattern were to provide the backbone of the locomotive equipment as regards the haulage of express trains well into the twentieth century.

The British 4–4–0 of this classic form, with inside cylinders and firebox dropped between the coupled wheels, had much to commend it. The deep firebox between the coupled axles yielded a high proportion of heating surface in relation to the grate area. The leading bogie made for easy riding over curves, and the distribution of weight, with about two-thirds available for adhesion, was satisfactory.

The cylinders and smokebox, placed centrally over the bogie, provided ample room longitudinally for the machinery and for a convenient length of boiler barrel. On cross section the design was the most compact imaginable. The cylinders between the frames delivered the power at the nearest point to the central line of the machine, and both inlet and exhaust arrangements were of the simplest and shortest possible character.

Credit for the introduction of this form of locomotive in Great Britain goes to Thomas Wheatley, who in 1871 brought out a series of engines which, save for the provision of a bogie, differed little from his previous 2–4–0s. They were small engines with 17 in. × 24 in. cylinders, 6 ft. 6 in. coupled wheels and boilers furnishing 1,059 sq. ft. total heating surface. Grate area was 17 sq. ft. and their total weight was just over 36 tons.

The second series of locomotives of this description was also of Scottish origin. Their designer, James Stirling, of the Glasgow & South Western, introduced them in 1873, and 22 were built during the ensuing four years. They bore unmistakable family characteristics, with cabs of the Stirling pattern, perforated splashers (favoured by his brother Patrick in his earlier work) and domeless boilers. The safety valves were mounted near the cab but were of the Ramsbottom uncovered type. 7 ft. 1 in. driving wheels were actuated by 18 in. × 26 in. cylinders. Boilers, pressed to 140 lb. per sq. in. and 4 ft. 2 in. diameter outside the smallest ring, contained 252 1½-in. tubes. The fireboxes furnished 84 sq. ft. heating surface and 16 sq. ft. grate area. Nearly 26 of the 39 tons total weight were available for adhesion. A notable feature of these engines was the short bogie wheelbase, 4 ft. 10 in. In contrast to later practice on some railways, the pivot was placed 1 in. ahead of the centre.

S. W. Johnson, then of the Great Eastern, was the first to produce an engine of this form in England. Two only were built, in 1874, and appeared shortly after their designer had left the Great Eastern for the Midland. Cylinders 17 in. × 24 in. drove 6 ft. 6 in. coupled wheels. Boilers, 4 ft. in diameter inside the smallest of the three courses of which the barrel was formed, contained 214 1¾-in. tubes. The fireboxes, 5 ft. 5 in. long, produced 100 sq. ft. heating surface. The grate area was 16·6 sq. ft. and the working pressure 140 lb. per sq. in. Some 26 of the 39¼ tons rested on the coupled wheels. These engines exhibited all their designer's distinction of outline. One, not perpetrated by him, was the raised running plate over the coupled wheels, and the form it took, so far as the writer can recall, was, in a 4–4–0, unique, the arc joining the higher and lower levels being described from a point corresponding with that of the wheel centre.

Johnson developed this form of engine in his subsequent career in the Midland, producing several classes of increasing capacity during his long tenure of office (Plates Nos. 10 and 11). A series of ten built by Kitson in 1876–7 had 17½ in. × 26 in. cylinders, 6 ft. 6 in. coupled wheels, 1,215 sq. ft. heating surface, 17·5 sq. ft. grate area, 140 lb. per sq. in. working pressure and boilers 4 ft. 2 in. in diameter. Johnson's 4–4–0s were of varying dimensions, including 5 ft. 6 in., 6 ft. 9 in. and 7 ft. coupled wheels. The boiler pressure was raised in the mid '80s to 160 lb. per sq. in. and the cylinder diameter enlarged to 18½ and 19 in. On the other hand, the heating surface was reduced by the employment of fewer tubes. The '1667' series of 1884 had 205 1¾-in. tubes producing, with the firebox (110 sq. ft.), a total heating surface of 1,122 sq. ft., and weighed 42¾ tons, of which nearly 28 rested on the coupled wheels. Small as they were, these engines performed admirably and with the single wheelers of the '90s and the later developments incorporating Belpaire fireboxes, to be referred to in a subsequent chapter, formed an important ingredient in the locomotive stock of the Midland.

Kirtley's excellent engines for the London, Chatham & Dover, introduced in 1877, had 17½ in. × 26 in. cylinders and 6 ft. 6 in. coupled wheels. 4 ft. 3 in. diameter barrels housed 200 1¾-in. tubes and fireboxes 5 ft. 9 in. long furnished 107 sq. ft. heating surface and a grate area of 16·3 sq. ft. The boilers were pressed to 140 lb. per sq. in. and nearly 28 of the total weight of 41¾ tons was available for adhesion.

James Stirling, whose Glasgow & South Western engines have already been described, introduced the same design on his transfer to the South Eastern. His first series—the 'Maidstones' of 1879—had 6 ft. coupled wheels, 18 in. × 26 in. cylinders, 4 ft. 4 in. (mean) diameter boilers, constructed with three rings, and 5 ft. 6 in. fireboxes.

A later series, introduced in 1883, consisting of 88 engines, the '240' class, had 7-ft. driving wheels (surprisingly large for the somewhat dilatory requirements of that railway), 19 in. × 26 in. cylinders, 140 lb. per sq. in. working pressure, 202 1⅝-in. tubes housed in 4 ft. 4 in. (maximum) diameter boilers and fireboxes 5 ft. 9 in. long, providing 103 sq. ft. heating surface and 16·8 sq. ft. grate area. Just over 28 tons of their total weight, 42½ tons, rested on the coupled wheels.

Barton Wright's 4–4–0s for the Lancashire & Yorkshire, introduced in 1880, had 17½ in. × 26 in. cylinders, 6-ft. coupled wheels, 1,054 sq. ft. heating surface, and weighed just over 42 tons.

McDonnell's engines of 1884 for the North Eastern, the first 4–4–0s with inside cylinders and frames on that railway were good engines, which suffered from the designer's tactlessness in dealing with the men accustomed to the easy-going, but basically far from lax, autocratic rule of Edward Fletcher. With their 17 in. × 24 in. cylinders, 6 ft. 8 in. coupled wheels, 4 ft. 2 in. diameter boilers, fireboxes producing 104 sq. ft. heating surface and 16·8 sq. ft. grate area and total weight of only 39½ tons, they were somewhat small for contemporary requirements, and were quickly superseded by the 2–4–0 'Tennants' already described.

The Great North of Scotland, which, as we shall see, originally favoured outside cylinders, introduced the inside type in 1884. These locomotives were the work of James Manson, who was responsible for some excellent locomotives on this line and, subsequently, on the Glasgow & South Western. The engines under notice had 17½ in. × 24 in. cylinders, 6-ft. coupled wheels, 189 1¾-in. tubes, 140 lb. per sq. in. boiler pressure and fireboxes furnishing 90 sq. ft. heating surface and 18 sq. ft. grate area. They weighed just over 37 tons, of which the bogie carried all but 13. These engines set the pattern for new construction on the Great North of Scotland for the rest of its career. They were used for passenger and goods work alike, the proportion of 4–4–0s to the rest of the locomotives stock being far higher on this railway than on any other in Great Britain.

Two notable Scottish examples of this pattern of engine were Drummond's North British and Caledonian engines of 1877 and 1884 (perpetuated in larger form on his subsequent removal to the London & South Western). Both had 18 in. × 26 in. cylinders and 6 ft. 6 in. coupled wheels. The former had a grate area of 21 sq. ft., a total heating surface of 1,099 sq. ft. and weighed 44 tons. Corresponding dimensions of the Caledonian engines were 19·5 sq. ft., 1,210 sq. ft. and 45 tons.

Charles Sacré's 4–4–0 for the Manchester, Sheffield & Lincolnshire, introduced in 1877, departed from the standard form in having outside frames for the coupled wheels, but not for the bogies. The external framing was carried forward, with deep slots to the buffer beam. Principal dimensions were: cylinders 17 in. × 26 in., coupled wheels diameter 6 ft. 3 in., boiler diameter 4 ft. 6 in., total heating surface 1,016 sq. ft. (firebox 94 sq. ft.), boiler pressure 140 lb. per sq. in., total weight 41 tons.

Fletcher's 'Whitby Bogies' of 1864 also had outside frames for the coupled wheels. With the smokebox placed forward of the bogie pin, they were designed for the sharp curves of the Whitby line and are typical neither of contemporary North Eastern nor general British practice.

Outside Cylinder 4–4–0s

Curiously enough, the 4–4–0 with outside cylinders made its first appearance in this country some years before the inside cylinder type, which is far more representative of British locomotive practice. As has already been pointed out, the employment of outside cylinders with the 4–4–0 wheel arrangement enables the front of the firebox to be brought a few inches nearer the leading coupled axle, with a corresponding increase in the length of the firebox set between the axles; but this consideration can hardly have influenced early designers, whose locomotives were relatively small. Given the preference for outside cylinders, however, the earlier replacement of a single leading pair of wheels by a bogie is readily explicable, since it provided a simple means of eliminating excessive overhang at the leading end.

Credit for the first 4–4–0 to run in this country goes to Bouch, of the Stockton & Darlington, whose engines with 6-ft. driving wheels were introduced in 1860, to be followed in the following year by others with 7-ft. wheels. Space need not, however, be occupied by a description of these somewhat crude machines, or of the 'Ginx Babies' with their 30-in. piston stroke, of 1871, none of them being typical of current or subsequent practice.

D. J. Clark's engines of 1861 for the Great North of Scotland were of more normal, though far from conventional, design. These had raised fireboxes with large domes above them, 16 in. × 22 in. cylinders, and 5-ft. coupled wheels. They were small engines, having a total heating surface of 956 sq. ft. and only 10¼ sq. ft. grate area. The design was developed by William Cowen, whose engines had 17 in. × 24 in. cylinders, 5 ft. 6 in. coupled wheels, boilers 4 ft. 4 in. in diameter, a total heating surface and grate area of 1,023 and 14 sq. ft. respectively and weighed nearly 39 tons, of which the bogie carried 12 tons. Outside cylinder 4–4–0s formed the backbone of G.N.S.R. locomotives stock for all kinds of traffic until the introduction of Manson's inside cylinder locomotives, to which reference has already been made.

The outside cylinder 4–4–0 made its first appearance on the neighbouring Highland in 1873, when David Jones substituted a bogie for the leading wheels on one of his 2–4–0s. A year later new 4–4–0s were built, retaining the Crewe type of frame, which afforded outside support for the cylinders, but not for the axles. Cylinders 17 in. × 24 in. drove 6 ft. 3 in. coupled wheels. Total heating surface and grate area were 1,228 and 16¼ sq. ft. respectively. The boilers were pressed to 140 lb. per sq. ft. and 14½ of the total weight of 41 tons were carried by the bogie.

Benjamin Connor's 'Dundee bogies' of 1877, for the Caledonian, provided a further example of the popularity of this form of locomotive in early Scottish practice. They were designed as 2–4–0s but delivered under Connor's successor as 4–4–0s, it is thought at the suggestion of Neilson's, their builders. We are still in the range of very small engines. The grates offered no more than 14·6 sq. ft. in area. The total heating surface was 987 sq. ft. Horizontal cylinders, 18 in. × 24 in., actuated large wheels, 7 ft. 2 in. in diameter. The boiler pressure was 130 lb. per sq. in. and the total weight was just under $41\frac{1}{2}$ tons.

The greatest exponent of the outside cylinder 4–4–0 was unquestionably William Adams of the Great Eastern and, subsequently, of the London & South Western.

The Great Eastern engines, 'The Ironclads', were considerably larger than any of those noted above; but it is understood they were not a success, and were ultimately relegated to goods work. They had 18 in. × 26 in. cylinders, 6 ft. 1 in. coupled wheels, 140 lb. per sq. in. working pressure, boilers 4 ft. 4 in. in diameter, containing 177 $1\frac{7}{8}$-in. tubes, and fireboxes 6 ft. long furnishing 100 sq. ft. heating surface and 17·3 sq. ft. grate area. Adhesion and total weight were just over 28 and 45 tons respectively.

On the South Western, Adams achieved real success. His were not the first 4–4–0s on that railway, W. G. Beattie having introduced the type in 1876, not, it must be said, with happy results. His engines had $18\frac{1}{2}$ in. × 26 in. cylinders, 6 ft. 7 in. coupled wheels, 4 ft. 2 in. boilers, 1,035 sq. ft. heating surface, grates affording 17·9 sq. ft. area, and a weight of $43\frac{3}{4}$ tons.

Adams's first 4–4–0s were the 'steam rollers' of 1879, mixed traffic engines with 5 ft. 7 in. wheels and solid bogie wheels, from which they probably derived their name. The fireboxes furnished 106 sq. ft. heating surface and 17 sq. ft. grate area. The boilers were pressed to 140 lb. per sq. in. and $29\frac{3}{4}$ of the total weight of just over 46 tons were available for adhesion.

His first express engines, of the following year, had 6 ft. 7 in. driving wheels, ten more square feet of heating surface and nearly 1 sq. ft. more grate area, the same boiler pressure, 234 $1\frac{3}{4}$-in. tubes, and a total weight of 46 tons 8 cwt. A further series of 6 ft. 7 in. engines, introduced in 1884, were preceded a year earlier by the first 7 ft. 1 in. class. These had identical boilers, 4 ft. 3 in. in diameter housing 218 $1\frac{3}{4}$-in. tubes, and fireboxes of the same dimensions as the earlier engines. The pressure was raised to 160 lb. per sq. in. and nearly 30 tons rested on the coupled wheels. The larger-wheeled engines weighed a trifle more than the others, 46 tons 2 cwt. against $45\frac{3}{4}$ tons. Developments of both these classes will be mentioned hereafter.

Mention should finally be made of some very similar but smaller engines supplied by Beyer Peacock (the builders of some of the South Western engines) to the Midland & Great Northern Joint Railway between 1881 and 1888. These had 17 in. × 24 in. cylinders, 6-ft. coupled wheels, 1,083 sq. ft. total heating surface and 17·7 sq. ft. grate area. They carried 140 lb. per sq. in. working pressure and weighed $38\frac{1}{4}$ tons.

IV SIX-COUPLED GOODS

The merits of the 0–6–0 wheel arrangement, combined with inside cylinders and a firebox placed between the second and third axles, have already been indicated and amply explain the extensive use made of this form of locomotive on almost all the railways of Great Britain.

To take the four largest concerns in our period—the London & North Western, the Midland, the Great Western and the North Eastern—all these, though their locomotive policies might differ in other respects, were at one in possessing large numbers of inside cylinder 0–6–0s of this conventional pattern. Outside frames were favoured by the Midland and the Great Western in the earlier years but eventually all came into line in placing the frames between the wheels.

'Long Boiler' Pattern

Before describing representative machines of this form, a few further examples should be given of the 'long-boiler' type in which all three axles were placed ahead of the firebox.

A good example is furnished by the Stockton & Darlington (later North Eastern) 'Panther' class, introduced in 1860, and built by Hawthorn Leslie, it was adopted as standard by William Booch, many of the engines being rebuilt in later years by Fletcher, McDonnell and Worsdell—an indication of the successful manner in which they performed their duties.

As built, they had 16 in. × 26 in. cylinders and 5-ft. coupled wheels. 6 ft. 2 in. separated the first and second coupled axles, the third axle being brought within 5 ft. 3 in. of the latter. Another characteristic of the design was the length of the boiler—13 ft. 7 in. 4 ft. 1 in. in diameter, it contained 191 2-in. tubes. The firebox, necessarily restricted by being placed behind the rear axle, was only 4 ft. 4 in. long, furnished 100 sq. ft. heating surface and a diminutive grate area of 11·8 sq. ft. The boiler was pressed to 120 lb. per sq. in. and the total weight was about 32¾ tons. Worsdell's boilers, 2 in. larger in diameter, contained fewer tubes (165). The firebox was some 4 in. longer, which resulted in an enhanced grate area of 13·3 sq. ft. These modifications increased the weight to 35 tons.

The Normal Type

In contrast with these, the Fletcher engines of the normal type (Plate No. 4), built in 1875, had 4 ft. 2 in. boilers containing 181 2-in. tubes and fireboxes providing 103 sq. ft. heating surface and 19·3 sq. ft. grate area. Their total heating surface, owing to the shorter barrel, was nearly 280 sq. ft. less than that of the Hawthorn Leslie engines (1,204 sq. ft. as against 1,483). Fletcher's engines weighed just over 36 tons.

T. W. Worsdell's two-cylinder compound engines of 1887 represent the highest stage of development to which the 0–6–0 was brought on the North Eastern during this period. High- and low-pressure cylinders were 18 and 26 in. in diameter respectively, both having a stroke of 24 in. Boilers, 4 ft. 3 in. in diameter, furnished a total heating surface of 1,136 sq. ft. of which 110 sq. ft. were provided by the firebox. The grate area was 17.2 sq. ft., the boiler pressure was 160 lb. per sq. in. and the total weight 43 tons 16 cwt.

The most extensive user of the 0–6–0 at the end of our period (and probably during the whole of it) was the Midland, which, like the North Eastern, handled an enormous goods traffic. The earlier examples were the work of William Kirtley, who favoured double frames, those inside terminating ahead of the firebox. They were built in several series and the dimensions varied. Those of 1869 had 17 in. × 24 in. cylinders, 5 ft. 2 in. coupled wheels, 118 2-in. tubes, firebox heating surface and grate area of 107 and 16·8 sq. ft. respectively, a boiler pressed to 140 lb. per sq. in. and a total weight of about 36 tons.

Johnson's engines were also produced in several series, with variations. Those built in 1878, which may be taken as representative, had 17½ in. × 26 in. cylinders and 5 ft. 2 in. wheels. 110 of the 1,313 sq. ft. total heating surface was provided by the firebox. Grate area and boiler pressure were 17½ sq. ft. and 140 lb. per sq. in. and the engines weighed 37 tons 18 cwt.

Before this period opened the London & North Western had an excellent class of 0–6–0s in Ramsbottom's famous 'DX' class already described. These served as a model for Webb's 'coal engines' of 1873–92, and his '18-in. goods' or 'Cauliflowers' introduced in 1880—both built in large numbers, the former class comprising 500 engines, the latter 310.

They exhibited in a marked degree Crewe's ability to turn out a serviceable engine at minimum cost; and, if the result was austere, it gave satisfaction, even delight, to the observer with an eye to the essentials of locomotive design and construction. Both classes were similar in many respects; the fireboxes furnished 94½ sq. ft. heating surface and 17·1 sq. ft. grate area, and both had a working pressure of 140 lb. per sq. in. and a boiler diameter of 4 ft. 2 in. The 18-in. goods had the advantage of 100 sq. ft. more tube heating surface. The coal engines were provided with 17 in. × 24 in. cylinders and 4 ft. 3 in. wheels, and weighed 29 tons 11 cwt. Corresponding 'Cauliflowers' dimensions were 18 in. × 24 in., 5 ft. 1 in. and 33 tons 7 cwt. In addition from 1881 on he and Webb rebuilt 500 of his predecessor's DX engines with boilers of similar dimensions to those fitted to his coal engines. These were known as 'Special DX'.

Like the Midland, the Great Western favoured double frames at the outset of this period and single frames at the end of it. Here also the series were of marked variety and the ground must be covered by referring to representative classes. Armstrong's engines, of which 310 were built between 1866 and 1876, had 17 in. × 24 in. cylinders and 5 ft. 2 in. wheels. The last double framed engines for the Great Western were the work of William Dean, built in 1885 and the following year. These had 17 in. × 26 in. cylinders, 5 ft. 1 in. wheels, a total heating surface and grate area of 1,157 and 15·2 sq. ft. respectively, 140 lb. per sq. in. working pressure and weighed 36 tons 18 cwts. The inside frame type attained its final development in the famous 'Dean goods' of the '90s.

Sacré's substantial engines of 1880 for the Manchester, Sheffield & Lincolnshire Railway may be taken as our last example of an outside framed 0–6–0. With their 17½ in. × 26 in. cylinders, 4 ft. 9 in. coupled wheels, boilers 4 ft. 5 in. in diameter and fireboxes providing 87·5 sq. ft. heating surface and 18¼ sq. ft. grate area and 40 tons weight, they were among the most powerful 0–6–0s of their day.

In order to keep this chapter within reasonable bounds our reference to the numerous classes of the standard 0–6–0s on other railways must be restricted to three examples—two of them being of exceptional capacity.

Patrick Stirling's engines of 1871, designed to cope with heavy coal traffic, had exceptionally large cylinders, 19 in. × 28 in., and 5 ft. 1 in. coupled wheels. 232 1¾-in. tubes and fireboxes furnishing 112 sq. ft. heating surface and 18·7 sq. ft. grate area were provided and the engines turned the scale at 40 tons. His standard goods engines had 17½ in. × 26 in. cylinders, 5 ft. 1 in. wheels, boilers 4 ft. 0½ in. in diameter outside the smallest ring containing 174 1¾-in. tubes, fireboxes furnishing heating surface and grate area of 92·4 sq. ft. and 16·25 sq. ft. respectively, 160 lb. working pressure and a total weight of 36½ tons (Plate No. 14).

Stroudley's engines for the London, Brighton & South Coast, of 1871, apart from their unusual size for those days, were remarkable in having the grates sloped upwards from the front in order to clear the rear axle. In the result a grate area of 19·3 sq. ft. was provided. The boilers of 4 ft. 6 in. maximum diameter contained 247 1¾-in. tubes. Cylinders 17½ in. × 26 in. drove 5 ft. wheels and the engines weighed 38 tons 12 cwt. It is noteworthy that engines of the same type introduced by Stroudley's successor many years later, were much smaller machines.

Locomotives of the same pattern were produced during this period by, among others, Dugald Drummond for the North British and Caledonian, by James Stirling for the Glasgow & South Western and the South Eastern, by J. A. F. Aspinall for the Lancashire & Yorkshire, by S. W. Johnson and his successors for the Great Eastern, by William Adams for the London & South Western (in reversal of his usual policy which favoured outside cylinders) and by William Kirtley for the London, Chatham & Dover.

Descriptions of these would take us beyond the space that should be devoted to this part of our subject. The examples already chosen are fully representative of the classic 0–6–0, which performed such excellent work of all kinds on British railways during the mid-Victorian period and for many years thereafter.

A notable departure from this classic design was made by William Adams of the Great Eastern, who, shortly before his departure for the South Western, prepared the general lines of a powerful 2–6–0 with outside cylinders. Modifications were introduced by Massey Bromley, his successor. 19 in. × 26 in. cylinders drove 4 ft. 10 in. coupled wheels. The boiler, of 4 ft. $6\frac{1}{8}$ in. maximum diameter, was pressed to 140 lb. per sq. in. and contained 240 $1\frac{3}{4}$-in. tubes. 102 sq. ft. heating surface and 17·8 sq. ft. grate area were provided by the firebox and the engines weighed just over $46\frac{1}{2}$ tons, all but $8\frac{1}{2}$ of which, borne by the pony tank, were available for adhesion.

It would be pleasant to record that this pioneer class of a form of locomotive destined to play a large part in British practice in later years was a success. But this was not so. These engines were heavy on coal and oil and all 15 of them, built in 1878, failed to survive ten years.

V TANK ENGINES

In Great Britain the tank engine has always been favoured. Large numbers of densely populated areas which resulted from the industrial revolution called for suburban services entailing relatively short runs and a quick turn-round, and an engine capable of being driven with equal facility in either direction. Moreover, so long as an engine of sufficient capacity could be provided, the accommodation for the coal and water on the same framework was no less advantageous. The demands for power of this description during the mid-Victorian era were not such as to require, for secondary work, the maximum development of boiler power in relation to permissible weight, while the gradual diminution of adhesion weight, with the progressive consumption of water and coal, was not, at this stage of locomotive development, of serious practical effect.

Tank engines introduced during these 30 years were of enormous number and variety, and the only way of keeping the present treatment within reasonable compass is to select some of the more interesting examples. It may, however, be noted at the outset that the typical British tank locomotive at the end of this period had inside cylinders, frames between the wheels and side tanks; and the somewhat cursory survey to which this section of our subject must be restricted will be designed to bring out these salient features which assert themselves with increasing emphasis in the course of the period.

In general we shall follow the plan adopted in our treatment of tender engines, dividing the subject into sections devoted to four-coupled and six-coupled engines, the first-named being further divided into those carried on three, four and (in one case) five axles.

In the 'six-wheeled four-coupled' class the carrying axle can, of course, be placed ahead of or behind the coupled wheels. On the whole the advantages would appear to lie with the latter arrangement; the heavier part of the locomotive, and that less affected by losses due to the consumption of coal and water, being available for adhesion, while the bunker space was less restricted.

2–4–0

The Great Western was a great user of both varieties, building of the 0–4–2 for branch service extending well into the present century. On the whole, Swindon inclined to the 2–4–0 and Wolverhampton to the 0–4–2 type. 2–4–0 tanks were much favoured for suburban work. Joseph Armstrong's 'Metro tanks' of this wheel arrangement were smart little engines with outside bearings for the leading wheels and inside cylinders 16 in. × 24 in. driving 5 ft. 2 in. coupled wheels. Boilers, 4 ft. 2 in. in diameter and pressed to 150 lb. per sq. in., housed 245 $1\frac{3}{4}$-in. tubes.

Firebox heating surface and grate area were 97·6 and 16·4 sq. ft. respectively. 1,100 gals. of water and just over two tons of coal were carried and the engines weighed a little under 44 tons, of which 31 tons 12 cwt. rested on the coupled wheels. These typical Great Western products, with cab roofs barely extending over the footplate and short open bunkers were much in evidence at Paddington and other busy centres of this company's territory.

Joseph Beattie's 2–4–0s for the neighbouring London & South Western, of which more than 80 were built between 1863 and 1875, with their horizontal outside cylinders and well tanks, were a very different proposition. The frames were inside throughout, though the auxiliary bearings for the leading wheels attached to the underside bars, already noted in connection with the same designer's 2–4–0 tender engines, were a feature of the tank engines also. Cylinders 15½ in. (later 16 and 16½) × 20 in. drove the leading pair of 5 ft. 6 in. coupled wheels. Diminutive boilers, only 3 ft. 8 in. in diameter, contained 224 1⅝-in. tubes, and the fireboxes provided a mere 80 sq. ft. heating surface and 14·8 sq. ft. grate area. Boiler pressure was 130 lb. per sq. in. and the engines weighed all but 34½ tons, of which practically 24 were available for adhesion. The well tanks were situated over the leading axle and under the footplate. Other distinctive features of these engines were their raised fireboxes, with domes over them, and feed water heater. A few of them survive to the present day.

A third no less distinctive class of 2–4–0 tank engine was that produced by F. W. Webb for the London & North Western in 1876. These engines exhibited the characteristic lines upon which the standard British tank engine was destined to develop—inside frames throughout, inside cylinders and side tanks. They were also noteworthy in being the first engines incorporating Webb's radial axle. Cylinders 17 in. × 20 in. actuated 4 ft. 7½ in. coupled wheels. The total heating surface was 972 sq. ft. and the boiler pressure was 130 lb. per sq. in. They carried 900 gals. of water and weighed 35¾ tons. More adequate fuel supplies were carried when an additional radial axle behind the coupled axles was provided in 1880 on the first of a series of 2–4–2 tanks which formed the standard North Western branch engine for many years.

0–4–2

The earliest 0–4–2 tanks calling for mention are those designed by McConnell for the Southern Division of the London & North Western in 1860. These had 15 in. × 22 in. cylinders, 5 ft. 6 in. coupled wheels, inside frames throughout and well tanks. In contrast, the London, Chatham & Dover 0–4–2 tanks, of 1866, had outside frames throughout and well tanks. The cylinders were 16½ in. × 22 in.; 4 ft. 1¼ in. boilers contained 183 2-in. tubes and just over 27 of the 41 ton 7 cwt. total weight rested on the coupled wheels.

On the Great Western, the first Wolverhampton side, as opposed to saddle, tank of this wheel arrangement appeared in 1870. These were quite small engines, even for their period, with 15 in. × 24 in. cylinders and 5-ft. coupled wheels. The frames were inside. 138 2-in. tubes were housed in boilers of 3 ft. 6 in. diameter inside. Firebox heating surface and grate area were 74 sq. ft. and 12·3 sq. ft. respectively. When full the side tank carried 620 gals. of water.

The most famous engines of this wheel arrangement were undoubtedly those designed by William Stroudley for the London, Brighton & South Coast, of which 125 were built between 1873 and 1887. These engines exhibited all the characteristics of that designer's work, already described with reference to his tender engines of the same wheel formation. The tank engines had 17 in. × 24 in. cylinders, 5 ft. 6 in. coupled wheels, boilers 4 ft. in diameter pressed to 150 lb. per sq. in. containing 175 1¾-in. (later 207 1⅝-in.) tubes and fireboxes 5 ft. 2¼ in. long providing heating surface and grate area of 91 and 15 sq. ft. respectively. They carried 1½ tons of coal and 860 gallons of water, and weighed 38½ tons. 27 tons rested on the coupled wheels. These engines, which coped with the branch line and suburban traffic of the Brighton line for many years, were an outstanding example of British locomotive engineering skill.

4–4–0

Many designers at this time combined the use of a bogie with four coupled wheels for tank engines; but here again opinions differed on the question whether the carrying wheels should lead or follow the drivers. In the long run the 0–4–4 proved to be the more favoured type, but several examples of the, perhaps superficially more attractive, alternative arrangement exist and these may be dealt with first.

London was the principal centre for 4–4–0 tank engines. On the North London Railway distinct varieties, both the work of William Adams, existed, one—the earlier—with inside cylinders and outside frames to the bogie, the other with outside cylinders in accordance with the plan favoured by that engineer in his later work on the Great Eastern and London & South Western. In each of the North London classes the cylinders were of the same dimensions, 17 in. × 24 in., so virtually was the grate area, 16½ sq. ft., and the boiler pressure, 160 lb. per sq. in., and the total weight, 44 tons. The earlier engine, introduced in 1863, had 5 ft. 9 in. coupled wheels, and 120 2¼-in. tubes, corresponding figures for the later type being 5 ft. 5 in., 186, and 1¾-in. Side tanks were provided and an unusual feature of the outside cylinder class, the first of which appeared in 1868 and which became the standard engine on this line, numbering 74 in all, was that the running plate was not carried forward beyond the front of the tanks. They were delightful little engines and furnish one of the not very numerous original designs which proved themselves in every way the equal of locomotives planned on more conventional lines.

Another original design, familiar to Londoners, was that of the Metropolitan and District 4–4–0 tanks, which were introduced in the '60s and numbered in all 120 engines. Here there was a marked similarity both in design and dimensions. The bogie wheel base was very short, the cylinders being placed above them and somewhat sharply inclined. In the earlier engines the four wheel truck was anchored to the frame 6 ft. 8 in. behind the centre line; but bogies were afterwards substituted. In each class the cylinders were 17 in. × 24 in. and the coupled wheels were 5 ft. 9 in. in diameter. Boilers 4 ft. 2 in. in diameter housed 2-in. tubes (164 in the District and 166 in the Metropolitan engines), and each carried 1,000 gals. of water and 1½ tons of coal and weighed just over 42 tons, of which 31 rested on the coupled wheels. The Metropolitan engines had the advantage of 11 sq. ft. in firebox heating over their District rivals, whose fireboxes provided 90 sq. ft. heating surface and 16 sq. ft. grate area.

For a 4–4–0 of conventional design, with inside cylinders, frames and side tanks, it is necessary to cross the Border to the North British, where Dugald Drummond produced two charming variations—one in 1879, with 17 in. × 24 in. cylinders and 6-ft. coupled wheels, the other a year later for light work with 16 in. × 22 in. cylinders and 5-ft. coupled wheels (Plate No. 7). The former had 220 1¾-in. tubes, the latter 142—figures which sufficiently indicate their difference in size.

0–4–4

The engines just mentioned provide a good introduction to the 0–4–4 of conventional pattern (Plate No. 6), with side tanks, inside cylinders and inside frames throughout. But it took time for this characteristic form of British locomotive to become standard practice; and a preliminary glance at some unusual types which it ultimately superseded will be appropriate.

Kirtley's 0–4–4 tanks for the Midland, introduced in 1869, provide an excellent example of earlier work. These engines had outside frames for the coupled wheels and inside frames for the trailing bogie. The water was carried in well tanks and the rear sand box was placed on the top of the firebox. They had 17 in. × 24 in. cylinders, 5 ft. 2 in. coupled wheels and boilers 4 ft. 2 in. in diameter housing 244 1⅝-in. tubes. Fireboxes, 5 ft. 6 in. long, provided 110 sq. ft. heating surface and 16·1 sq. ft. grate area. Quite substantial engines for their time, but of an unusual character. Like most of Kirtley's creations they had a very long life and were a familiar sight in the London district long after the Midland had ceased to exist.

Fletcher's 0–4–4 locomotives for the North Eastern, introduced in 1874, provide another example of well tank engines, in their case with inside frames. Cylinders 16 in. × 22 in. drove 5-ft. coupled wheels. Total heating surface and boiler pressure were 1,163 sq. ft. and 140 lb. per sq. in. respectively, and the engines weighed 43½ tons. Succeeding engines of the same design were built to larger dimensions.

A still more unusual design of 0–4–4 tanks was that designed by Benjamin Connor for the Caledonian in that these engines were provided with outside cylinders—an awkward arrangement, as we have seen, when combined with leading coupled wheels. They had 17 in. × 22 in. cylinders, 4 ft. 8 in. coupled wheels and weighed nearly 43¼ tons of which 26 were available for adhesion.

Cudworth's 0–4–4 tanks for the South Eastern, of 1866, presented a complete contrast with inside cylinders (15 in. × 20 in.) and outside frames throughout. The coupled wheels were 5 ft. 7 in. in diameter, and the total heating surface was 903 sq. ft. They were small engines weighing only 33¾ tons of which the coupled wheels carried 22 tons.

We conclude this part of our subject by giving three examples of the typical British 0–4–4 tank with inside cylinders and frames and side tanks.

Johnson's Midland engines, introduced in 1883, had 18 in. × 24 in. cylinders, 5 ft. 4 in. coupled wheels, 110 sq. ft. firebox heating surface, 16 sq. ft. grate area, carried just over 2 tons of coal and 1,270 gals. of water, and weighed 53¼ tons, 43 of which rested on the coupled wheels (Plate No. 12).

Adams's South Western engines of 1888–1896, of which 50 were built, had boilers standard with the 0–4–2 'Jubilees' above noted. Cylinders 18 in. × 26 in. drove 5 ft. 7 in. coupled wheels. The boiler pressure was 160 lb. per sq. in. and just over 35 of the total weight, 53 tons, were available for adhesion.

Patrick Stirling's 0–4–4 tanks, which in the latest, and largest, form, were introduced in 1889, had 18 in. × 26 in. cylinders and 5 ft. 7½ in. coupled wheels. Boilers, 4 ft. 2½ in. in diameter, pressed to 160 lb. per sq. in., housed 174 1¾-in. tubes. Fireboxes, 5 ft. 6 in. long, furnished 92½ sq. ft. heating surface and 16¼ sq. ft. grate area. They carried 1,000 gals. of water and weighed all but 53½ tons of which about 20¼ was carried by the bogie. A noteworthy feature of these engines was the lenth of wheelbase between the rear coupled and the leading bogie wheels—10 ft. 3 in.; this, though usually to a less marked extent, being characteristic of 0–4–4 tank engines and reflecting the freedom accorded to designers in providing an adequate firebox with this form of locomotive.

2–4–2

A third way of supporting a four-coupled, four-axle locomotive is, of course, to provide a single pair of carrying wheels at either end. This plan, which was in effect a tank engine version of the 2–4–0 tender passenger locomotive, found favour with a number of designers, the firebox normally being placed between the coupled axles.

Robert Sinclair's 2–4–2 well tanks for the Great Eastern were not of this pattern, the firebox lying behind the second coupled axle. This entailed the provision of a long boiler (13 ft. 6 in.). Pressed to 120 lb. per sq. in. and of only 3 ft. 11 in. in diameter, it contained 143 $1\frac{7}{8}$-in. tubes. The firebox, of necessarily restricted length (4 ft. 6 in.), furnished 69 sq. ft. heating surface and a meagre grate area of 11·7 sq. ft., and less than 19 tons of the total weight of $36\frac{1}{4}$ tons rested on the coupled wheels. They were introduced in 1864. Twenty were built.

T. W. Worsdell's engines, the first of which appeared twenty years later, were of the conventional pattern. 18 in. × 24 in. cylinders drove 5 ft. 4 in. coupled wheels. Boilers, 4 ft. 2 in. in diameter and pressed to 140 lb. per sq. in., contained 198 $1\frac{3}{4}$-in. tubes, and the firebox heating surface and grate area were 98·4 and 15·4 sq. ft. respectively. They weighed all but 52 tons with just over 29 carried by the coupled wheels.

The most renowned engines of this wheel arrangement were those of the Lancashire & Yorkshire. Introduced by Aspinall just before the end of our period and perpetuated by his successors, this company's 2–4–2 tanks performed work which might well have seemed impossible for engines of such relatively small dimensions. Aspinall's locomotives had 18 in. (some $17\frac{1}{2}$ in.) × 26 in. cylinders and 5 ft. 8 in. coupled wheels. The boilers, pressed to 160 lb. per sq. in., housed 220 $1\frac{3}{4}$-in. tubes, and the fireboxes provided 108 sq. ft. heating surface and $18\frac{3}{4}$ sq. ft. grate area. They weighed all but 56 tons of which 33 tons 13 cwt. were available for adhesion. Passing reference has already been made to Webb's engines of this wheel formation for the North Western.

4–4–2

A natural development of this form of locomotive was derived from the substitution of a bogie for the single axle of the leading end. During the mid-Victorian period all the engines of this description, so far as the writer's recollection goes, had outside cylinders save that designed by T. H. Riches for the Taff Vale Railway and built in 1888. Cylinders $17\frac{1}{2}$ in. × 26 in. drove 5 ft. 3 in. coupled wheels. The boilers, pressed to 160 lb. per sq. in., contained 214 $1\frac{3}{4}$-in. tubes and the grate area was 19 sq. ft. 1,600 gals. of water and $2\frac{1}{4}$ tons of coal were carried and 31 of the 54 tons total weight were available for adhesion.

The pioneer of the outside cylinder 4–4–2 tank was the London, Tilbury & Southend, which used it as the standard type throughout its existence (Plate No. 18). The first of these engines, whose design has been authoritatively attributed to William Adams, then in charge of the Great Eastern locomotive department at Stratford, appeared in 1880. Their 6 ft. 1 in. coupled wheels —driven by 17 in. × 26 in. cylinders—rendered them suitable for the fastest trains on the line. 4 ft. 2 in. diameter boilers contained 200 $1\frac{5}{8}$-in. tubes, and fireboxes, 6 ft. long, provided a heating surface of 97 sq. ft. and 17·2 sq. ft. grate area. Water and coal capacity were 1,300 gals. and 2 tons and the total weight was 56 tons, 2 cwt. with 33 tons available for adhesion.

Unlike the Tilbury engines, which were genuine 'express tanks', Adams's South Western engines were planned for secondary—suburban and branch—duties. These attractive and unusual locomotives, which had no side tanks and long bunkers supported by the trailing axles, were practically a smaller version of the same designer's outside cylinder express engines. They had $17\frac{1}{2}$ in. × 24 in. cylinders, 5 ft. 7 in. coupled wheels and 201 $1\frac{3}{4}$-in. tubes housed in boilers 4 ft. 2 in. in diameter. Fireboxes, 6 ft. 2 in. long, furnished 111 sq. ft. heating surface and 18·1 sq. ft. grate area. The boiler pressure in the later engines was raised from 140 to 160 lb. per sq. in., and the tank capacity was increased from 1,000 to 1,200 gals. Nearly 31 of the total weight of 54 tons 2 cwt. rested on the coupled wheels. Of this class 71 engines were built in 1882–85. Following the introduction of his 'Jubilees', Adams abandoned the ten-wheeled tank in favour of the above described 0–4–4s. With these engines we conclude our survey of mid-Victorian four-coupled tank engines and turn our attention to the six-coupled variety.

0–6–0

The six-wheeled coupled tank locomotive, without carrying wheels, either in front or behind, was one of the most widely used designs during this period—and, of course, beyond it. In many respects it closely resembled the 0–6–0 tender engine. Power was imparted at the central point, the firebox occupied a position between the second and third axles and the whole of the weight —including in this case that of the water and fuel—was available for adhesion. In nearly every case inside cylinders were employed on the main line railways but there was not the same agreement about the position of the frames or the tanks. Outside frames, favoured in many of the earlier engines, tended to give way to the more usual arrangment. Saddle and side tanks were common during the whole of our period, the former facilitated access to the motion, apt to be obstructed by side tanks, but raised the centre of gravity (not a serious matter for engines employed chiefly on shunting and slow goods duties).

A detailed study of the 0–6–0 tank engines put on the road by designers between 1860 and 1890 would be of considerable interest. But nothing of the kind can be attempted here. We must content ourselves with giving single examples of the varieties above described.

Our first example—of a double-framed, inside cylinder side tank—may appropriately be taken from the Great Western, which of all the railways was the most extensive user of tank engines having this wheel formation. The engines selected for mention were built at Swindon in 1870. Cylinders 17 in. × 24 in. drove 4 ft. 6 in. wheels. 180 2-in. tubes were housed in barrels 4 ft. 2 in. in diameter. The fireboxes provided 98 sq. ft. heating surface and 16·85 sq. ft. grate area. The side tank held 850 gals. and the total weight was 37 tons 12 cwt. Short cabs were substituted for weather boards a few years after their reconstruction. These engines, though genuine Great Western creations, were hardly typical of that railway's practice, which for this class of locomotive favoured the saddle tank—subsequently to be developed into this company's unique 'pannier' tanks (Plate No. 41).

Nevertheless we propose to take the reader to the Rhymney Railway of South Wales for our example of an outside framed saddle tank. The first batch of these was built by Sharp Stewart & Company in 1872. 4 ft. 7 in. coupled wheels were driven by 16 in. × 24 in. cylinders. The boilers, 4 ft. 2 in. in diameter and pressed to 140 lb. per sq. in., contained 170 2-in. tubes. Firebox heating surface was 88·3 sq. ft. and the tanks held 1,000 gals. Weather boards at the boiler and end of the bunker were connected by a roof, without side protection.

The London, Brighton & South Coast provides in its E goods tank an excellent example of a side tank with inside frames. These engines, of which 72 were built from 1874 onwards, were of typical Stroudley design. They had 17 in. × 24 in. cylinders, 4 ft. 6 in. coupled wheels, 4 ft. diameter boilers, pressed to 150 lb. per sq. in., housing 160 $1\frac{3}{4}$-in. (later 231 $1\frac{1}{2}$-in.) tubes; fireboxes, 5 ft. $2\frac{3}{4}$ in. long, providing 87 sq. ft. heating surface and 15 sq. ft. grate area, coal and water capacity of $1\frac{3}{4}$ tons and 900 gals., and weighed $39\frac{1}{2}$ tons.

For an example of an inside framed saddle tank we may take Patrick Stirling's engines, the first of which appeared in 1868. These had $17\frac{1}{2}$ in. × 24 in. cylinders, 5 ft. 1 in. coupled wheels, a boiler of 3 ft. 9 in. diameter containing only 90 2-in. tubes, fireboxes 4 ft. 7 in. long and a tank capacity of 975 gals. In this, his first series of 0–6–0 tanks, the cab roof did not extend over the bunker.

Outside cylinder 0–6–0 tanks (apart from industrial locomotives) were relatively uncommon, and again we may enlist the Rhymney railway to give us an example of a saddle tank. These engines, built by the Vulcan Foundry in 1861, had 17 in. × 24 in. cylinders and 4 ft. 3 in. coupled wheels. The boilers, 4 ft. 1 in. in diameter and pressed to 140 lb. per sq. in., contained 173 2-in. tubes. The fireboxes provided 76 sq. ft. heating surface and 14·6 sq. ft. grate area. No cabs were provided. Contrary to the usual practice the saddle tanks, which held 1,080 gals., were carried forward over the smokebox. These engines were subsequently rebuilt with cabs and shorter saddle tanks leaving the smokebox free.

A more modern design of outside cylinder 0–6–0 tank locomotives was that produced by Park for the North London in 1879. These neat engines, intended for the company's goods traffic had horizontal cylinders and high side tanks merging with the cabs, the roofs of which extended to the back of the bunker. The cylinders were 17 in. × 24 in. and the wheels 4 ft. 4 in. in diameter. The grate area was $16\frac{1}{3}$ sq. ft. and the total heating surface was 895 sq. ft. of which the firebox provided 81. 880 gals. of water and $1\frac{1}{4}$ tons of coal were carried and the engines weighed practically 44 tons.

0–6–2

This period witnessed the introduction of the 0–6–2 type of tank engine, and although this wheel arrangement was not extensively favoured by designers at this time, it subsequently became very popular. The earliest example is furnished by Barton Wright's rebuilding in 1879 of a 0–6–0 goods engine with side tanks and lengthened frames to take a coal bunker, supported by a radial axle beneath it.

Two years later F. W. Webb introduced his 0–6–2 tank coal engines, of which 300 were built. This was in effect a tank version of the contemporary 0–6–0 goods engines and carried the North Western standard boiler of the period, which had a maximum diameter of 4 ft. 2 in. and a grate area of 17·1 sq. ft. Cylinders were 17 in. × 24 in. and coupled wheels 4 ft. $5\frac{1}{2}$ in. in diameter. They carried 1,150 gals. of water and 2 tons of coal, and weighed $43\frac{3}{4}$ tons.

Barton Wright's first 0–6–2 tanks built for the Lancashire & Yorkshire were passenger engines with $17\frac{1}{2}$ in. × 26 in. cylinders and 5 ft. 1 in. coupled wheels, and were larger than the North Western engines, weighing some 6 tons more. Ultimately this railway favoured the four coupled 'double ender' for passenger service, as we have seen.

0–8–0

Finally, mention should be made of two remarkable eight-coupled tank locomotives built for the Great Northern by the Avonside Engine Company of Bristol in 1866. These had outside cylinders ($18\frac{1}{2}$ in. × 24 in.) driving the third pair of 4 ft. 6 in. coupled wheels, and long side tanks. The boiler barrel, 13 ft. 8 in. long and 4 ft. 4 in. in diameter, contained 184 $2\frac{1}{8}$-in. tubes and produced 1,450 of the 1,550 sq. ft. total heating surface. They weighed 56 tons and had a comparatively short life, being broken up in 1880. Compare with these mammoths the diminutive 2–2–2 tanks employed at one stage on some lines (Plate No. 1).

It would not be right to conclude this review of mid-Victorian tank locomotives without emphasizing that we have been concerned substantially with the main lines of standard practice. Several cases of unusual designs of considerable interest have gone unrecorded; only so could the review have been kept within a reasonable compass and a balanced account of locomotive practice within these important thirty years be presented.

CHAPTER III

Late Victoriana

THE concluding years of Victoria's reign constitute a notable period in the development of British locomotive practice. When it opened, locomotives of quite moderate power, embodying no particularly original features compared with their predecessors sufficed; before it closed the increasing weight of trains compelled designers to turn their attention to more ambitious experiments, including the provision of additional axles to carry the larger boilers.

Then, as throughout the whole period of steam locomotive history, the problem of providing machines of adequate capacity presented itself in its more acute form in relating to the haulage of express trains; and it is to locomotives intended for these duties that attention will be chiefly directed in this chapter. For this purpose reliance in the main was placed on the 4-4-0 with inside cylinders and a deep firebox, dropped between the two coupled axles. The excellence of this general layout has already been indicated.

LATE VICTORIAN 4-4-0s

A typical 4-4-0 of the '90s would have 18 in. × 26 in. cylinders, 6 ft. 9 in. driving wheels, a boiler 4 ft. 4 in. to 4 ft. 8 in. in diameter, tubes 10 ft. to 11 ft. long, and a firebox up to 7 ft. in length, furnishing a grate area in the neighbourhood of 20 sq. ft. Such an engine would turn the scale at 50 to 55 tons, of which some 35 would rest on the coupled wheels. (The basic arrangement of such an engine is illustrated in the diagram sketch on page 49.)

New designs of engines of this description during the '90s were provided by Henry Ivatt for the Great Northern, S. W. Johnson for the Midland, Pollit and Parker for the Manchester, Sheffield & Lincolnshire, J. A. F. Aspinall for the Lancashire & Yorkshire; south of the Thames by Dugald Drummond for the London & South Western (founded on his earlier work for the North British and Caledonian), William Kirtley for the London, Chatham & Dover, James Stirling for the South Eastern, and Billinton for the London, Brighton & South Coast; and, in Scotland, by Peter Drummond for the Highland, Holmes for the North British, James Manson for the Glasgow & South Western and Pickersgill for the Great North of Scotland (Plates Nos. 13 and 16). This type was developed to a notable extent on the Caledonian, their designer, McIntosh, being generally regarded as the pioneer in the provision of large boilers.

13. Pickersgill's express locomotive for the Great North of Scotland, 1895
14. Goods engine of the Stirling-Ivatt period for the G.N.R.

15. Britain's first Atlantic by Ivatt, 1898
16 & 17. South Eastern 4–4–0's by James Stirling and Wainwright, 1898 and 1901

18. Express tank engine for the London, Tilbury and Southend Railway by Adams, 1897
19 & 20. Robinson's first 4–6–0 and eight-coupled goods engines for the Great Central, 1902

21 & 22. Johnson's Smith three-cylinder compound 1901 and L.M.S. version following the 1906 Deeley modificatio

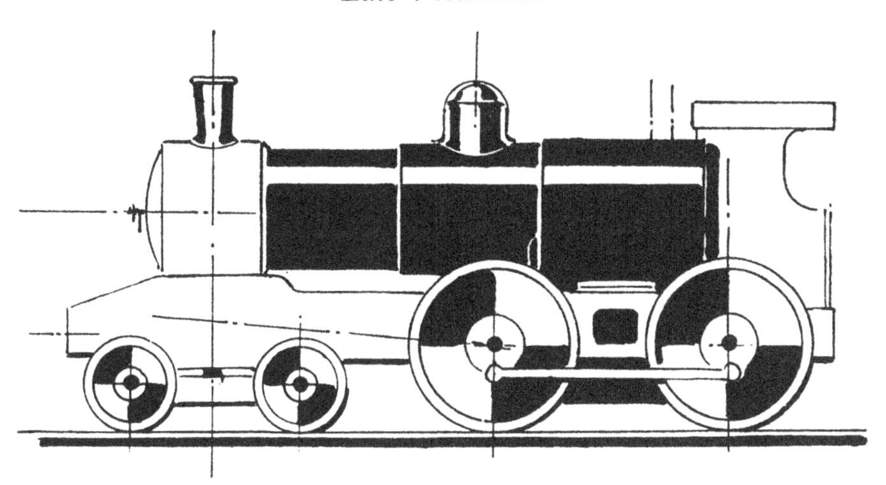

British 4–4–0. The General Plan

A New Departure

Of the Caledonian engines Charles Rous-Marten in the 100th issue of the *Railway Magazine* (October 1905) wrote: 'Those engines recognized for the first time the principle . . . that an express engine with 6 ft. 6 in. coupled wheels may have a boiler of sufficient internal diameter, 4 ft. 8 in., to make it bulge out laterally above those 6 ft. 6 in. wheels. Mr McIntosh is justly entitled to be named the leader of the movement which set in after the London–Aberdeen Race, of 1895, in favour of larger boiler-power.'

But perhaps this design reached its highest stage of development at this time in Wilson Worsdell's class 'R' locomotive, of 1899, for the North Eastern. The principal dimensions of these engines are worth recalling. Cylinders 19 in. × 26 in., with piston valves below them, drove 6 ft. 10 in. coupled wheels, which carried 35 tons 5 cwt. of the total weight of 51 tons 13 cwt. The maximum boiler diameter was 4 ft. 9 in. and the length between tube plates 11 ft. 10 in. A firebox 7 ft. long, and of good depth, was dropped between the coupled axles, set 9 ft. 6 ins. apart, and provided 144 sq. ft. heating surface and 20 sq. ft. grate area. These engines must be ranked among the most brilliant of our period, and, if it were necessary to select among North Eastern classes produced in substantial numbers the outstanding product, class 'R' would be the most likely candidate for the highest honours.

Variations on the theme are provided by the employment of outside frames, outside cylinders and the principle of double expansion.

In this respect, as in many others, the Great Western pursued a line of its own in locomotive matters. Negatively expressed, its individuality is nowhere more remarkable than in the fact that it—and it alone among all but a few minor railways—never produced a 4–4–0 of the conventional pattern. The deviation from standard in this case was in the provision of outside frames, both for bogie and coupled wheels. The inside frames, supporting the cylinders and smokebox, extended from the buffer beam to the front of the firebox and furnished additional bearings for the driving axle, which was thus supported at four points—a valuable feature in large 4–4–0 engines with inside cylinders, which restricted the dimensions of the inside bearings, upon which, of course, in the conventional plan, exclusive reliance was placed. On the other hand the design entailed the employment of cranks to impart the motion of the driving to the rear coupled wheels.

This form of locomotive achieved its ultimate development, in late Victorian times, in the

5 ft. 8 in. 'Camel' and the 6 ft. 8 in. 'Atbara' classes, in which the above described design of frame was combined with a boiler no less unusual. A Belpaire firebox was raised high above the parallel barrel and provided very ample steam space in the area of maximum ebullition. Steam was taken from the front corners of the firebox, and a dome was dispensed with, its customary place on the barrel being occupied by the safety valves. These highly unconventional features (for the period) are reflected in the rugged, uncompromising appearance of the engines.

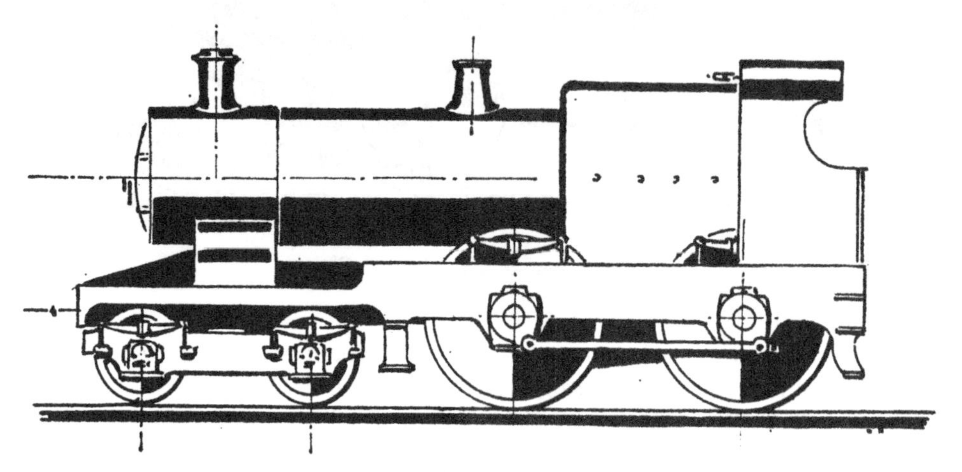

A Great Western Atbara

Outside Cylinders

The next variation is concerned with the position of the cylinders. Only Adams, on the London & South Western, and David Jones, on the Highland, favoured the outside position for 4–4–0s during this period, and before it ended the plan had been abandoned by their successors. The South Western engines were remarkably graceful machines, particularly the last two batches produced in the '90s, one with 6 ft. 9 in. wheels for use west of Salisbury, the other with 7 ft. 1 in. wheels for the more level road east of that point. An advantage in the provision of outside cylinders was the additional space afforded for the firebox between the coupled axles by the absence of a crank axle, which with inside cylinders had, of course, to be set back to clear it. This point is well illustrated by a comparison between these Adams engines and their Drummond successors of 1898. The coupled wheelbase and length of firebox of the former were 9 ft. and 6 ft. 10 in. respectively, the corresponding dimensions of the latter were 10 ft. and 7 ft. 4 in. For their period these final Adams creations were large machines. The 7 ft. 1 in. engines had boilers of 4 ft. 4 in. maximum diameter, pressed to 175 lb. per sq. in., the firebox providing 122 sq. ft. heating surface and 19·6 grate area, and they weighed just over 50 tons.

The Highland engines broke new ground on that railway, since, for the first time, the Crewe type outside framing—which had been an invariable feature of this company's 4–4–0 locomotives —was abandoned in favour of inside frames. With their 19 in. × 24 in. cylinders, 6 ft. 3 in. coupled wheels, boilers pressed to 175 lb. per sq. in. and 4 ft. $4\frac{7}{8}$ in. in diameter, fireboxes providing 119 sq. ft. and 20·5 sq. ft. grate area, and practically 30 of a total of $47\frac{1}{2}$ tons available for adhesion, they were admirably suited to Highland conditions. The fact that the last was built in 1917, long after they had been superseded as a class, by inside-cylinder locomotives, may be regarded as a tribute to their excellence.

L.S.W. Express Locomotive by Adams

Compounds

The third variation consisted in the employment of double expansion. Compounding was extensively employed on the North Western and the North Eastern. In the former case it survived the Victorian era, in the latter it did not.

F. W. Webb's 'Jubilees', which derived their name from Queen Victoria's diamond jubilee in 1897, were the first 4–4–0s to be built for the London & North Western and differed markedly from their eight-wheeled predecessors—the 'Greater Britains' and 'John Hicks'—also products of the '90s—in which carrying wheels were placed fore and aft the uncoupled drivers. The 'Jubilees' had four cylinders, all driving the same axle (the two earlier classes had three, and divided drive, the outside high-pressure cylinders actuating the second, and inside high-pressure the first driving axle). Only two sets of valve gear were used for the four valves, those for the outside high-pressure cylinders deriving their motion through rocking shafts. This was a source of weakness, since the high and low pressure cut-offs could not be varied independently.

The high and low pressure cylinder proportions were also unsatisfactory. With a common stroke of 24 in., the 'Jubilee' had 15 in. high-pressure and 20½ in. (originally 19½ in.) low-pressure cylinders—a proportion of 1 to 1·87. They were, moreover, hopelessly under-boilered and their performance was mediocre. The larger 'Alfreds', the first of which appeared a few months after the close of our period, in their altered form, with additional valve gear for the high-pressure cylinders, enabling the cut-off to be varied as between high- and low-pressure cylinders, did well—sometimes remarkably well—but this alteration was never carried out on the 'Jubilees'.

The North Eastern favoured two-cylinder compounds, but shortly after the opening of this period all the 4–4–0s of this description were converted to simples, save one; and it is with this engine, No. 1619, that we are now concerned. The reconstruction was in accordance with the system devised by W. M. Smith, Chief Draftsman at Darlington. Steam (at 200 lb. per sq. in.), taken by a single high-pressure cylinder (19 in. × 26 in.), was exhausted in the steam chest action of the two external, low-pressure (20 in. × 24 in.) cylinders. The crank pins of the latter were set as in an ordinary two-cylinder engine—at 90 deg. to each other—and the low-pressure crank bisected the obtuse angle between them. The cylinders were placed in line below the smokebox and drove the leading pair of wheels. Pressure in the low-pressure steam chest could be augmented up to a predetermined maximum by means of a reducing valve under the driver's control. No provision was made for exhausting the high-pressure steam direct to the atmosphere, and the engine could therefore be operated as a compound or a semi-compound, but not as a simple. The Smith system was the converse of the Webb, in which the high-pressure engine was duplicated, and, with the three cylinders of not widely differing diameter, certainly looked better on the drawing-board than the latter, in which the high-pressure cylinders were diminutive and the low-pressure cylinder enormous. Moreover, the splitting of the exhaust between two low-pressure cylinders was easier on the fire. On the other hand, in those days—and for long afterwards—when the greater part of the work of a compound was performed by the high-pressure engine, the Webb system, in terms of power exerted by the pistons, was preferable—a fact which makes it tantalizing to speculate what would have been the result of the employment of this arrangement, unhampered by the defects from which the Webb three-cylinder compounds suffered. However that may be, there can be no question of the merits of the Smith system which, with minor modifications, was destined to be used to an incomparably greater extent than any other in this country. Indeed it made its mark as the one successful compound system standing to the credit of British engineers.

THE SINGLE-WHEELER

A substantial proportion of major express work during our period was performed by single-wheelers. Here it would be unrealistic to draw the distinction between the late Victorian period and that which preceded it too rigidly. On the Great Northern, for example, the express trains during the '90s, were hauled principally by the previously described Stirling 4–2–2 'eight-footers' and the same designer's 2–2–2s, dating from 1870 and 1868, though additions of modified design were made to each class during the '90s. In 1900 and the following year, they were joined by the Ivatt 4–2–2s, which were substantially an enlarged bogie version of the Stirling 2–2–2s. These Ivatt singles, the last singles to be built for work in this country, had 19 in. × 26 in. cylinders, 7 ft. 7½ in. driving wheels, and weighed 48 tons 11 cwt., of which 17 tons 15 cwt. were available for adhesion. The boiler, pressed to 175 lb. per sq. in. was of substantial size for an engine of this description. The maximum diameter was 4 ft. 5 in. and the length between tube plates 11 ft. 8 in. The firebox, 7 ft. long provided 126 sq. ft. heating surface and 23·2 sq. ft. grate area.

The single-wheeler was also much in evidence on the neighbouring Midland. The design originated in 1887 but as the great majority of the 95 locomotives were constructed during our period and most of them belonged to the second, third, and fourth series, which were of modified design, the whole class may be treated as falling within the scope of this chapter. The size of the coupled wheels was progressively enlarged from 7 ft. 4½ in. in the first batch (28 engines) to 7 ft. 6½ in. in the second (52 engines), and 7 ft. 9 in. and 7 ft. 9½ in. (25 engines) in the third and fourth series. The last named had 19½ in. × 26 in. cylinders and a working pressure of 180 lb. The boiler was of 4 ft. 3 in. maximum diameter and 10 ft. 6 in. in length. The firebox, 8 ft. long, furnished 147 sq. ft. heating surface and 24·5 sq. ft. grate area. Adhesion and total weights were, respectively, 18 tons 10 cwt., and 50 tons 3 cwt.

Holden's 4–2–2s, built for the Great Eastern in 1898, very similar to the Midland engines though of less graceful outline, had but a short career on front rank work, their deficiency in adhesion weight being doubtless responsible for their removal to lighter duties on the advent of the 'Claude Hamiltons' only two years later. The principal dimensions of these 4–2–2s were: cylinders 18 in. × 26 in., coupled wheels 7 ft. diameter, maximum boiler diameter 4 ft. 4 in., length between tube plates 11 ft. 4 in., firebox heating surface 114 sq. ft., grate area 21·3 sq. ft., boiler pressure 160 lb. per sq. in., weight in working order 49 tons (adhesion 19 tons).

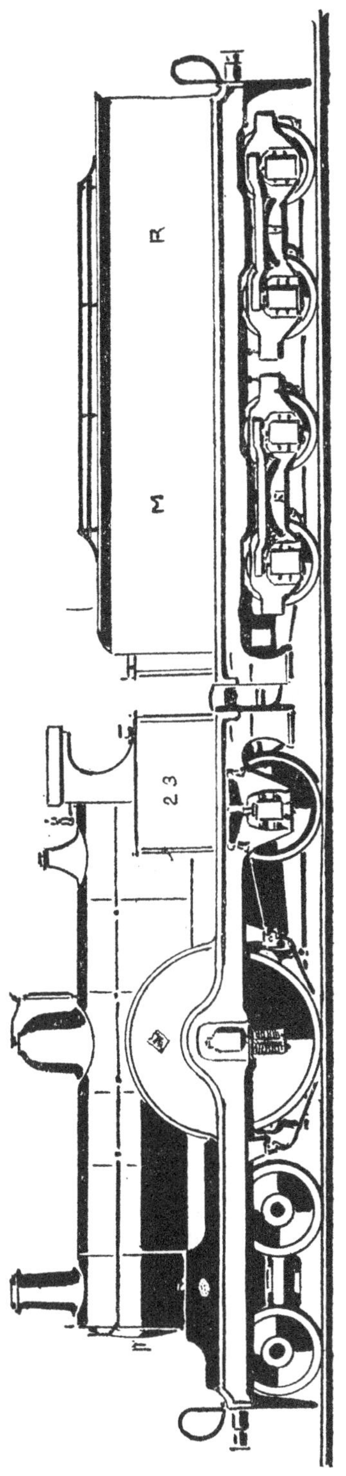

A Midland Single by Johnson

Pollitt's 4–2–2s for the Great Central also had a comparatively short career on the London extension, for which they were designed, but performed excellent service for many years on the Cheshire Lines Committee's metals on the Liverpool–Manchester service, provided under the aegis of the Great Central, Great Northern and Midland railways. With their 19½ in. × 26 in. cylinders, 7 ft. 9 in. driving wheels, 4 ft. 3 in. diameter boilers and fireboxes providing 142 sq. ft. heating surface and 24·8 sq. ft. grate area, 200 lb. per sq. in. boiler pressure and 18 tons 7 cwt. of their 47 tons 5 cwt. total weight available for adhesion, these engines were, for their type, of considerable power.

The North Eastern 4–2–2s, converted by Wilson Worsdell to simples within this period, but constructed before it as two-cylinder compounds by his brother T. W. Worsdell, played a far less prominent part in the haulage of express trains than did the singles of the southern partner of the East Coast route to Scotland.

The Great Western, on the other hand, like the Great Northern and the Midland, was much addicted to the single-wheeler in the '90s and many of the best expresses out of Paddington were hauled by the fine series of 80 engines designed by William Dean and all built during this period. They were a modified version of a series of 2–2–2s, built in 1891–92, which, following a derailment, were subsequently fitted with bogies.

These locomotives had 19 in. × 24 in. cylinders and 7 ft. 8½ in. driving wheels, and boilers 4 ft. 3 in. maximum diameter and 11 ft. 6 in. between tube plates. Steam was taken from a very capacious dome but the fireboxes were raised, though of circular form at the top. These, 6 ft. 4 in. long, furnished 127 sq. ft. heating surface and 20·8 sq. ft. grate area. The boiler pressure was 160 lb. per sq. in., and 18 tons of the total weight, 49 tons 11 cwt., was borne by the driving wheels.

None of the railways south of the Thames or north of the Border constructed any single-wheelers during this period.

All these 'singles' had inside cylinders and, broadly, conformed to the same general pattern. But, in the position of the frames, they exhibited an extraordinary variety. Outside frames had, from the inception of the steam locomotives, been favoured by many designers, fortified in many instances by frames in the usual position for the axles subjected to the greatest strains, which were provided with four bearings. This plan was adopted for the driving axles by Johnson on the Midland, Dean on the Great Western and Holden on the Great Eastern, all of whom also provided outside bearings for the trailing wheels. Dean furnished outside bearings for the bogies as well. At the opposite extreme Pollitt's engines had inside frames throughout. So had Ivatt's and Worsdell's, save for the trailing wheels, which were provided with external frames and axle boxes. (It may be recalled that Stirling's 4–2–2s had outside cylinders and inside bearings throughout, his 2–2–2s had inside bearings for the driving and outside bearings for the carrying wheels, while the little Ramsbottom 2–2–2s, still much in evidence as pilots on the North Western, resembled the Stirling 'eight-footers' in having outside cylinders and inside frames throughout.)

In general, the dimensions of these single-wheelers, when compared with those of the 4–4–0s, reflect both the limitations and potentialities of the type. The diameter of the boilers was restricted by the distance between the tyres, which imposed a maximum capable of being accommodated between them. On the other hand, the presence of a trailing, in place of a second coupled, axle gave the designer greater freedom in the provision of an adequate firebox, of which he was not slow to take advantage, as a comparison of the areas of the respective grates will show. Both these are, of course, vital dimensions in relation to operative capacity; but the principal limiting factor of the single-wheeler lay in the small proportion of the weight capable of being utilized for adhesion, and the increasing weight of trains rendered the type unsuitable for the hardest work with the dawn of the twentieth century, notwithstanding the invention of steam sanding which, indeed, alone gave it a renewed span of life during the '90s.

GOODS AND TANK LOCOMOTIVES

The goods engines of comparable capacity were of the 0–6–0 type. Engines of this description, with inside frames and inside cylinders, were to be found on all the large British railways, even the Great Western, for once, falling into line, and, indeed, in the excellent 'Dean Goods' dating from 1883, providing one of the best examples. For the heaviest work, however, that railway depended upon 2–6–0 inside-cylinder locomotives with double frames, a goods counterpart of the 'Atbara' 4–4–0s, to which reference has already been made. Considerations of space preclude further reference to the ubiquitous 0–6–0 which, with the opportunities offered for the provision of a well-proportioned boiler and firebox and compact machinery operating at the lateral and longitudinal centres of the mechanism and with every ounce of weight available for adhesion, constituted one of the most admirable steam power-units ever conceived and one, moreover, highly representative of all that was best in British locomotive practice (Plate No. 14).

Nor need space be occupied with the tank locomotives, whether of the 2–4–2, 0–4–4, 0–6–0 or 0–6–2 orders, used for lesser traffic; for these locomotives presented no problems of design which did not arise in a more acute form in the production of the larger engines. A passing reference should, however, be made to the Whitelegg, Adams-inspired, outside-cylinder 4–4–2 tanks of the London, Tilbury & Southend Railway, which were virtually express engines, and carried out with marked success the exacting duties required of them (Plate No. 18).

SIGNIFICANT DEVELOPMENTS

The concluding part of this chapter will be devoted to the development which took place (largely towards the end of this period) to meet increasing traffic demands.

0–8–0

The first departure from existing standards was made by F. W. Webb, and consisted of adding a further axle to the customary three provided on locomotives designed for goods traffic. The first of these 0–8–0 engines, built in 1892, was provided with inside cylinders actuating the second coupled axle, through somewhat short connecting rods. The second engine was a three-cylinder compound with the high-pressure cylinders outside, the second axle being driven by all three. Comparative tests were made between the two in which—of course!—the compound proved superior and this was taken as the prototype for many others.

These machines were of moderate dimensions. The wheels were equally spaced over a distance of 17 ft. 3 in. The boilers, of 4 ft. 3 in. diameter, were 15 ft. 6 in. long, the front tube plate being recessed some 2 ft. into the barrel. The grate area was 20·5 sq. ft. and the boiler pressure 175 lb. per sq. in. They weighed 49¼ tons.

The ineradicable sluggishness of a locomotive compounded on the Webb system was of little account on engines destined for mineral traffic, and their four-cylinder successors of 1901 may, perhaps, be regarded as that designer's most successful efforts with double expansion.

4–6–0

The second break-away from contemporary practice is due to the enterprise of David Jones, of the Highland, whose 4–6–0s, Nos. 103–117 of 1894 (the first of them still happily with us), were the pioneers of a type destined to be more widely used than any other for engines employed for heavy duties.

The new locomotives constituted an important departure from Highland practice, their predecessors having been provided with double frames of the 'Crewe' pattern. They had the normal inside frames and outside cylinders driving the middle coupled axle, which was 7 ft. 9 in. distant from the third, sufficient to accommodate a firebox of adequate depth and of the same length. The throat plate was brought as near as possible to the driving axle; and the depth of the firebox was 5 ft. 4¾ in. below the boiler centre line at this point and sloped up 1 ft. 2½ in. at the back, which lay just behind the third coupled axle. This was a familiar plan for 4–6–0s, though in no other instance were these axles separated so far apart in relation to the length of the firebox. The first and second coupled axles, were placed as close as possible, only 5 ft. 6 in. apart. The engines were designed for goods traffic over the mountainous gradients of the Highland Railway and had 20 in. × 26 in. cylinders and 5 ft. 3 in. coupled wheels. The boiler was 4 ft. 7⅞ in. in diameter and 13 ft. 9 in. long. The firebox furnished 113·5 sq. ft. heating surface and 22·6 sq. ft. grate area. The boiler was pressed to 175 lb. per sq. in. and the engine turned the scale at 56 tons of which 42 were available for adhesion. For their period these engines were of great power.

Peter Drummond's 'Castles', introduced in 1900, were a highly successful enlargement of that design. These engines, planned for express work, had 19½ in. × 26 in. cylinders, 5 ft. 9 in. coupled wheels, boilers 5 ft. in diameter and fireboxes 8 ft. long, furnishing 26·5 sq. ft. grate area and 134 of the 2,050 sq. ft. total heating surface. Boiler pressure was 180 lb. per sq. in. and the engines weighed 58 tons 17 cwt., of which nearly 44 tons rested on the coupled wheels (Plate No. 23).

In 1896 William Dean produced a somewhat ungainly 4–6–0 for the Great Western with inside cylinder and outside frames and a wide firebox above them. This was followed three years later, by a larger machine, but the design was not perpetuated and had no influence upon British locomotive practice.

To Wilson Worsdell of the North Eastern belongs the credit of producing the first 4–6–0 in this country for the haulage of express trains. No. 2001 of 1899 was provided with a boiler 4. ft. 9 in. in diameter and 15 ft. long. Outside cylinders, 20 in. × 26 in., drove the second pair of the 6 ft. 1¼ in. coupled wheels, which were equally spaced over a wheelbase of 14 ft. The firebox, 8 ft. long, was, as in the Highland engines, brought as close as possible to the driving axle but was somewhat shallow—4 ft. 9 in. below the boiler centre at the front and 3 ft. 9 in. at the back—owing to the presence of the third coupled axle beneath it. Firebox heating surface and grate area were 130 and 23 sq. ft. respectively. 49½ of the 66 tons were available for adhesion.

After a few years these engines were relegated to fast goods and excursion work, being supplanted on express trains by engines having 6 ft. 8¼ in. coupled wheels. The boiler was of the same diameter as that of the earlier engines but 10½ in. longer and the coupled axles were each 7 in. further apart. The fireboxes were of the same size. They weighed just over a ton more and carried about 3 tons more over the coupled wheels. Only five of these were built and Charles Rous-Marten used to express surprise that the class, which gave him the best work he experienced on the North Eastern, was never multiplied. They are also noteworthy as being the first 4–6–0s

N.E.R. Pioneer British Express 4–6–0

in this country to be provided with coupled wheels of virtually undiminished size and may therefore be regarded as the pioneer genuine six-coupled express engines.

'Atlantics'

The last, and in one sense most significant, development of this period to be recorded was the introduction in the country of the 'Atlantic' type. The limitation of the classic British 4–4–0 plan lay in the restriction placed on the length of the firebox by the distance separating the coupled axles, between which it lay, the limiting factor being the practical length of the coupling rods. Drummond, who had extended this to 10 feet—a dimension later to be equalled but never exceeded in British practice—obtained further length for the firebox by resorting to four cylinder propulsion and divided uncoupled drive in the No. 720 of 1897. The distance between the driving axles in this machine was 12 ft. and the firebox provided 27·4 sq. ft. grate area.

Apart from such an expedient as this, the only way to lengthen the firebox (and maintain the customary depth) was to place it behind the coupled axles, and this is what H. A. Ivatt of the Great Northern and John Aspinall of the Lancashire & Yorkshire did in their 'Atlantic' engines, introduced in 1898 and 1899 respectively. In the former the firebox was 8 ft. long, had a grate area of 26·75 sq. ft. and furnished 140 sq. ft. heating surface. In the latter, corresponding dimensions were 8 ft. 1 in., 26 and 175·8 sq. ft. Ivatt's engines worked at 175 lb. per sq. in. Aspinall's at 180.

The maximum boiler diameters were 4 ft. 8 in. and 4 ft. 9 in. respectively and the distance between the tube plates 13 ft. and 15 ft., the front tube plate in each case being recessed into the barrel. The Lancashire & Yorkshire engines were rather larger and had Belpaire fireboxes; otherwise the boilers were of similar design.

Here the resemblance ceased. Ivatt employed ($18\frac{3}{4}$ in. $\times$ 24 in.) outside cylinders driving the rear coupled axle and cut down to a minimum the coupled wheelbase and that between the bogie and the coupled wheels. With inside cylinders Aspinall used the first coupled wheels as the drivers and, notwithstanding a bogie wheelbase of only 5 ft. 6 in., his engine was longer by 3 ft. $3\frac{1}{4}$ in. between the leading bogie and rear coupled axles. The Great Northern engine had 6 ft. $7\frac{1}{2}$ in. coupled wheels carrying 31 of a total weight of 58 tons. The 7 ft. 3 in. wheels of the Lancashire & Yorkshire engine carried 35 of a total weight of $58\frac{3}{4}$ tons.

So began a new chapter in British locomotive history.

CHAPTER IV

The Broad Gauge

THE Broad Gauge, which was adopted by the Great Western from its inception in 1835 and was to prove something of an embarrassment for years prior to its abolition in 1892, owed its existence to the initiative of Isambard Kingdom Brunel—one of the most determined and famous railway engineers of all time. The problems of locomotive design in relation to a track 7 ft. 0½ in. wide were very different from those associated with the standard gauge of 4 ft. 8½ in., but it cannot be said that the greater freedom afforded by the wider track was fully exploited. Broad gauge engines showed some advantage over their standard gauge contemporaries, particularly in the larger grates capable of being placed between the frames; but on the whole the difference in capacity between the two was not as marked as might have been expected. This may well be due to the fact that the developments in locomotive design, in response to traffic demands, were capable of being met by machines built within the limits of the narrower track; but it is impossible to conduct, as is proposed, a brief review of broad gauge locomotive practice without a sense of disappointment that greater advantage was not taken of the opportunities presented.

Our review falls into four periods—preliminary, and those over which Gooch, Armstrong and Dean presided.

Brunel left the design of locomotives in the earliest period to the builders. They were an extremely mixed lot and only one of them, *North Star* (broad gauge engines were named and not numbered), can be pronounced a success. This was a 2–2–2 express locomotive with inside cylinders and outside sandwich frames built of wood bounded by iron plates—a form to which the Great Western was much addicted and which provided a useful measure of resilience on the unyielding road with its longitudinal timber supports under the rails and sleepers placed at greater intervals than those on the standard track. Additional frames were provided between the wheels and, in some cases, a central frame also—five in all.

Gooch

Sir Daniel Gooch's 2–2–2s of 1840, of which more than 60 were built, were a development of *North Star*. They had 15 in. × 18 in. cylinders, boilers 4 ft. in diameter containing 131 2-in.

tubes, and fireboxes of gothic contours, 4 ft. 6 in. long and 4 ft. 8½ in. wide, furnishing 91 sq. ft. heating surface and 12·6 sq. ft. grate area. They weighed 24¼ tons, of which just over 11½ rested on the driving wheels.

A further step was taken in *Great Western*—said, though this is disputed, to be the first product of Swindon works—which appeared in 1846. The 18 in. × 24 in. cylinders drove 8 ft. wheels. The boiler barrel, containing 278 2-in. tubes, and the firebox, 5 ft. 6 in. long and furnishing 151 sq. ft. heating surface and 22·6 sq. ft. grate area, showed a great advance on Gooch's earlier engines. The boiler was pressed to 100 lb. per sq. in. A fracture of the leading axle led to its replacement by a pair, which, being fixed to the main frames, were not a bogie. Nevertheless, it would be pedantic not to describe the rebuilt engine as a 4–2–2, which was to remain the standard type for express engines on the broad gauge during the remainder of its existence. In its altered form *Great Western* ran until 1870. A new *Great Western* was built in 1888.

The same general plan was adopted in the 'Iron Duke' class, save that the gothic firebox was replaced by one of the raised pattern, following the contour of the boiler barrel. The diameter of the latter, 4 ft. 9¾ in., was considerably more than that of any narrow gauge engine of the period and the cross-sectional measurement of the firebox, 6 ft., demonstrates the advantages of the broad gauge, even at the stage of development reached in the late '40s. With vertical sides this distance apart and a length of 5 ft., a grate area of nearly 22 sq. ft. and a firebox heating surface of all but 148 sq. ft. was provided. The barrel contained 303 2-in. tubes and the boiler pressure was 100 (later 115) lb. per sq. in. Only about 12¼ of the total weight of 35½ tons were available for adhesion. A noteworthy feature of these engines was the five-bearing driving axle, an arrangement rendered possible by the system of framing previously mentioned.

The six engines of this class were joined by 16 others of modified design (including the famous *Lord of the Isles*) in 1848–51 and by seven more a few years later. The outer firebox dimensions of the modified engines were the same as in the earlier series but, for some reason which the writer is unable to explain, the heating surface and grate area were increased to 156 and 25·5 sq. ft. respectively. The former dimension of the last series has been given as 171. The boiler pressure was raised to 120 lb. per sq. in. Eight of these locomotives survived until the abolition of the broad gauge, together with more than a dozen replacements, which took the former engines' names.

Gooch's 0–6–0 goods engines, of which there were 138, differed widely from their express counterparts, in that none of them, save the first four, had outside frames. The series built in 1847, which were larger than their predecessors, had boilers with raised round-topped fireboxes of the type fitted to the later 4–2–2s. The 16 in. × 24 in. cylinders drove 5-ft. wheels. Boilers, 4 ft. 3 in. in diameter, contained 219 2-in. tubes and fireboxes 4 ft. 11 in. long provided 121 sq. ft. heating surface and 18·4 sq. ft. grate area. The boiler was pressed to 115 lb. per sq. in. and the total weight was 28 tons 3 cwt. A later series, built between 1852 and 1856, had larger boilers of a standard pattern, 3 in. greater in diameter and containing 30 more tubes. The grates were fractionally larger (19·2 sq. ft.) and the weight was increased by a few tons.

Inside frames were also employed by Gooch in a series of 4–4–0 express engines introduced in 1855. These engines, which were intended to deal with heavier loads than the 'singles' but at the same speed, had 17 in. × 24 in. cylinders and 7-ft. coupled wheels. The carrying wheels were attached to the main frames, and the boilers—with raised round-topped fireboxes—were of the same pattern and dimensions as those fitted to the later 0–6–0s. The coupled wheels carried 21½ of the 36½ tons at which these engines turned the scale.

Gooch was also responsible for some 2–4–0s of generally similar design to that of the engines just described but smaller, as was appropriate for the secondary traffic for which they were intended.

Space does not permit of several other of this designer's products to be dealt with, but brief mention must be made of his 4–4–0 saddle-tanks for the lines beyond Bristol and his 2–4–0 'Metro' tanks.

The former, fitted as they were with bogies at the leading end, were genuine 4–4–0s. The frames were inside. They had 17 in. × 24 in. cylinders and, save for the first two, 5 ft. 9 in. coupled wheels. Here again the boiler was of the size and pattern fitted to the later 0–6–0s. They weighed about $38\frac{1}{2}$ tons. The saddle-tanks carried 930 gals. of water. Coal capacity must have been severely restricted by the shovel-shaped bunker which overhung the rear coupled axle.

The 'Metro' tanks were remarkable in being one of the few broad-gauge classes to be fitted with outside cylinders. These were steeply inclined and occupied a position ahead of the leading axle. Diameter and stroke were respectively 16 in. and 24 in., and they actuated the leading pair of 6 ft. coupled wheels—somewhat large, perhaps, for the Bishop's Road–Farrington Street section of the Metropolitan Railway for which they were intended. The boiler was 4 ft. 9 in. in diameter, and the firebox, 6 ft. long, provided a grate area of 18 sq. ft. Some 700 gals. of water were carried in the well-tanks below the barrel (the presence of which may explain the employment of outside cylinders) and the exiguous coal bunker. They were ungainly engines in marked contrast to Gooch's better work.

Pearson

At this point mention should be made of one of the most remarkable types of locomotive ever to run on British metals. These are the 4–2–4 express tank engines built to the designs of J. Pearson, in 1853, for the Bristol & Exeter Railway—a broad gauge line incorporated into the Great Western in 1876. Enormous driving wheels, 9 ft. in diameter, actuated by $16\frac{1}{2}$ in. × 24 in. cylinders, were flanked fore and aft by bogies having 4-ft. wheels and a wheelbase of 5 ft. 9 in. and set with their centres 9 ft. 6 in. distant from the vertical line passing through the driving axle. The boilers, fractionally more than 4 ft. in diameter, contained 180 $1\frac{5}{16}$-in. tubes, and the engines are said to have weighed about 42 tons. Some years later, the eight engines of which this batch consisted were replaced by four others of the same type but of modified design. The cylinder diameter was enlarged to 18 in. and that of the driving wheels reduced to 8 ft. 10 in. The raised fireboxes provided 146 sq. ft. heating surface and a grate area of 23·1 sq. ft., and the replacements were about $7\frac{3}{4}$ tons heavier than their predecessors.

Armstrong

Joseph Armstrong's tenure of office on the Great Western (to which we now return) lasted from 1864 to 1877. His work for the broad gauge included the production of another class of 2–4–0 passenger engines with inside cylinders and frames, somewhat smaller than Gooch's engines already mentioned, but of more modern appearance, incorporating, as they did, cabs of a pattern familiar on the Great Western down to the end of the Churchward regime nearly 80 years later. The curving of the running plate over the wheels and the absence of splashers represented a continuance of former practice, which was to last to the end of the broad gauge. These engines had 16 in. × 24 in. cylinders, 6-ft. coupled wheels, boilers 4 ft. in diameter containing 186 2-in. tubes, and solid plate frames in place of those of the sandwich pattern. 26 were built in 1865 and 1866, a few being subsequently rebuilt as tank engines with smaller wheels and larger cylinders.

A number of the 'Iron Duke' class were rebuilt during Armstrong's superintendency, with new boilers pressed to 140 lb. per sq. in., 4 ft. 9 in. barrels, housing 377 1½-in. tubes and fireboxes providing 164 sq. ft. heating surface. Others were replaced, on the assumption of office by William Dean in 1877, by new engines of similar design, which had 4 ft. 6 in. diameter boilers containing 330 (in some cases 332) 1⅝-in. tubes, the firebox heating surface being 132 sq. ft. in the earlier and 153 sq. ft. in the later replacements. A third series was remarkable in being provided with boilers 4 ft. 10 in. in diameter containing 375 1¾-in. tubes, which, with the 137 sq. ft. furnished by the firebox, gave the formidable total heating surface, for the period, of 2,085 sq. ft. So strengthened, the majority of the 'Iron Duke' class lasted until the abolition of the broad gauge.

Dean

With the advent of Dean we enter upon a new phase of broad gauge locomotive practice—that of the 'converted' and the 'convertible'. The end of the broad gauge was in sight and the task which fell to the locomotive chief was to make temporary provision for its needs.

This he did by furnishing some standard gauge engines with longer axles to enable them to work on the broad gauge—the 'converted'—and by ensuring that new broad gauge engines should be of such design as to be readily adaptable to the standard gauge—the 'convertible'.

Dean's first engines for the broad gauge were of the former category, comprising as they did a number of his standard 0–6–0s and 0–6–0 saddle-tank engines, which were fitted with longer axles.

The first of the convertibles was a series of ten 2–4–0 side-tank engines of typical Great Western form, built contemporaneously with a series of ten others for the standard gauge. The former had outside bearings for the leading wheels but, save for the abandonment of these bearings, conversion was effected by the simple process of shortening the axles and bringing the wheels within, instead of beyond, the outer frames.

The same plan was adopted with the 2–2–2 express engines introduced in 1891, eight of a batch of 30 being constructed to run on the broad gauge. The wheels of these engines were outside both sets of double frames, and of the others between them. The driving wheels had four bearings, the carrying wheels two, on the outer frames. They were typical Dean products, with very large domes and raised fireboxes. Large, 20 in. × 24 in., cylinders drove the 7 ft. 8½ in. wheels. The barrel housed 245 1¾-in. tubes and the firebox, 6 ft. 4 in. long and, of course, of the restricted width demanded by the standard gauge, provided 124 sq. ft. heating surface and 20·8 sq. ft. grate area. The driving wheels carried 19 tons of the total weight, which amounted to 43 tons 8 cwt. After conversion these engines and their standard gauge counterparts were fitted with leading bogies and formed, with others of similar design, one of the most famous classes of express locomotive ever to run on the Great Western or indeed any other railway in Britain.

CHAPTER V

Edwardian Engines

KING EDWARD VII reigned from January 22, 1901, until his death on May 6, 1910. This period witnessed a marked advance in British locomotive practice. Conditions were favourable. General prosperity was reflected in increased traffic on the railways, both passenger and goods. Competition spurred the various companies to give their best, and the fact that each was responsible for its own locomotives led to a combination of general advance and individuality in design which renders the period one of outstanding interest.

With the advance from the use of four axles to five—and in one instance to six—in the case of express engines, and from six coupled wheels to eight in the case of those intended for the heaviest goods traffic, the central problem confronting designers arose from the necessity to combine the Stephenson boiler, which demands a good depth of firebox for efficient combustion, with a disposition of wheels providing at once sufficient flexibility and adequate adhesion.

When our period opened, the heaviest express trains were, with a few exceptions, being hauled by locomotives of the 4–4–0 type. When it closed 4–4–2 or 4–6–0 type locomotives were in general use. A corresponding advance—from the 0–6–0 to the 0–8–0 and 2–8–0 types—is to be found in locomotives engaged in goods work.

4–4–0

In general our basic problem was not acute in relation to the classic British types with which the period opens (though, as we saw in chapter 3, the break-away had begun). In the 4–4–0 the firebox could be dropped between the coupled axles; in the 0–6–0 the same expedient was available and in any event the axles were sufficiently low to enable the grate to be placed over them without appreciable loss of depth.

The 4–4–0 with deep firebox between the coupled axles reached the highest point of development it was ever to attain within the Edwardian period. This occurred in 1904 when George Whale produced his famous 'Precursors' on the London & North Western Railway. Here for the first time a boiler exceeding five feet in diameter was fitted to the 4–4–0 type of engine; the maximum diameter was, in fact, 5 ft. 2 in. and it housed no less than 309 1⅞-in. tubes. The coupled axles were 10 ft. apart and this provided room for a firebox 7 ft. 4 in. long, and allowed the horizontal grate to be placed 6 ft. 3 in. below the boiler centre line. The result was a large firebox heating surface of 161·3 sq. ft.—a figure which exceeded that provided on the majority of the 'Atlantic' and 4–6–0 locomotives of the period and doubtless accounted in large measure for the phenomenal work which these 4-4-0 engines performed. Another factor contributing to the result may have been the excellent steam distribution effected by the Joy valve-gear, although its adoption was dictated by necessity rather than choice, the central longitudinal bearing with which these, in common with other North Western engines, were provided precluding the use of eccentrics.

Notable examples of this pattern of locomotive were introduced by J. Holden on the Great Eastern Railway in 1900, by J. G. Robinson on the Great Central in 1902, by D. Drummond on the London & South Western in 1903, by J. F. McIntosh on the Caledonian in 1904, and by H. Wainwright on the South Eastern & Chatham in 1901 (Plate No. 8) and 1905—all these being of considerably increased capacity compared with their predecessors on the same railways.

There is no locomotive design in which at some points clearances do not have to be reduced to a minimum so that full advantage may be taken of the facilities it affords; in none is further advance not barred by the limits thus imposed. The principal limiting factor in the design just considered was, as we saw in chapter 3, the restriction placed on the length of the firebox by the coupled wheelbase, which was in turn restricted by the practical length of the coupling rods. These, as has been stated, were 10 ft. long in Whale's locomotives. Drummond used rods of the same length, but no other designer went so far. (The position would have been eased by outside cylinders, which by eliminating the crank axle, would have enabled the firebox to be brought further forward; but this, of course, would have involved a substantial departure from the classic pattern.)

A simple expedient for enlarging the grate without lengthening the coupling rods was that adopted by S. W. Johnson on the Midland Railway, just before our period, in his first 'Belpaires'. Within the period there was a larger series of the same class and also his three-cylinder compounds (Plate No. 2). The grate sloped upwards so as to clear the rear coupled axle, a grate area of 25 sq. ft. in the 'Belpaires' and of 26 sq. ft. in the compounds being combined with a coupled wheelbase of 9 ft. 6 in. R. M. Deeley's compounds of 1905 and the corresponding simple engines, the '990s' of 1907, had 28·4 sq. ft. of grate area, without any lengthening of the coupling rods (Plate No. 22). This plan somewhat restricted the depth of the firebox, as may be illustrated by the fact that the 9 ft. fireboxes on the Deeley engines yield 152·8 sq. ft. heating surface, nearly 10 sq. ft. less than those of the 'Precursors', which were 1 ft. 8 in. shorter.

The sloping grate extending over the rear axle was adopted on the Derby engines built for the Somerset & Dorset Joint Railway (small as they were), by R. J. Billinton in 1901 for his large 4-4-0 s on the London, Brighton & South Coast (reversing his earlier practice in the 'Grasshoppers' of 1895), by James Manson in 1907 for engines of quite moderate dimensions on the Glasgow & South Western, and by Wilson Worsdell on the North Eastern in 1908 on his 'R1s', which, with their 5 ft. 6 in. diameter boilers, 27 sq. ft. of grate area and 42 tons over the coupled wheels, represented the highest stage of development to which the 4-4-0 type of express engine was brought during the Edwardian era. The grate area was the same as that of the corresponding 'Atlantics', but the firebox was shallower as it had to clear the coupled axle beneath it, and the limitation imposed by its presence may be illustrated by the fact that with an identical length of firebox the 4-4-0s had 158 and the 'Atlantics' 180 sq. ft. of heating surface.

The Great Western, as always, was a law unto itself. In his 'Cities' of 1903 G. J. Churchward combined the double frames reminiscent of an earlier period of British locomotive practice, much favoured on this railway, with a boiler remarkable alike in form and lateral dimensions. The double frames enabled four bearings to be fitted for the driving axle—a good feature in high-powered inside cylinder engines. The firebox extended back over the trailing coupled axle, which was placed only 8 ft. 6 in. from its counterpart. The coned barrel widened from 4 ft. $10\frac{3}{4}$ in. at the smokebox to 5 ft. 6 in. at the point where it joined the firebox casing, which was itself 3 in. wider in front at the centre line and brought in to 4 ft. 9 in. at the rear. The boiler was thus largest at its most effective point for steam raising, and it is of interest to note that the dimension of 5 ft. 9 in. at the front of the firebox casing was unequalled during this period in the productions of any other designer so far as express engines were concerned, whether of the 4–4–0, 4–4–2 or 4–6–0 order. Otherwise, with its 11-ft. barrel, 7-ft. firebox and 20·5 sq. ft. grate area, the boiler employed on the 'Cities' was of substantial, though by no means remarkable, size for the period.

In the following year Churchward used the same boiler in conjunction with inside frames and outside cylinders of the size and pattern adopted for his 4–6–0 and 'Atlantic' engines. This position of the cylinders was unique for a 4–4–0 engine, so far as new construction during the Edwardian era is concerned. It was favoured, however, by the unprecedented piston stroke of 30 in. (so avoiding a crank axle with a 15-in. throw) and was also in accordance with the system of standardization which was so conspicuous a feature of Great Western locomotive practice, then as thereafter. The conventional British design of 4–4–0 was, as we have seen, never adopted by this railway, and the foregoing are two interesting variations on the theme.

THE 'ATLANTIC' TYPE

Another expedient for increasing the capacity of the firebox was to place it behind the coupled wheels and to provide an additional pair of wheels to carry it, giving a 4–4–2 wheel arrangement. This plan which removed the restriction, both as regards length and depth, had already been adopted by H. A. Ivatt in the Great Northern and J. A. F. Aspinall on the Lancashire & Yorkshire Railway in their 'Atlantics' of 1898 and 1899, described in chapter 3. The former, it will be recalled, adopted outside cylinders driving the second pair of coupled wheels, the latter inside cylinders driving the first pair. Our period witnessed considerable developments of the Ivatt plan; the Aspinall plan remained unique, though some multi-cylinder designs with the front pair of wheels as the drivers may be regarded as derivatives.

It is at this stage that we witness the principal revolution (apart from increase in size) that took place in British locomotive practice during the Edwardian period—the changeover from inside to outside cylinders. The outside position enabled the second pair of coupled wheels to be used as the drivers and the distance between the rear bogie and leading coupled axle to be kept to the minimum permitted by the size of the wheels themselves. But this advantage, which, of course, had a corresponding beneficial effect on the total wheelbase, was purchased at what must be regarded as the reluctant abandonment of the time-honoured cylinder position which had become all but universal in British practice.

The Ivatt plan was adopted by Robinson on the Great Central in his beautiful 'Atlantics' of 1903, by Wilson Worsdell on the North Eastern and W. P. Reid on the North British in their massive machines of 1903 and 1906, and by Churchward in *Albion*, converted from a 4–6–0 in 1904 to afford comparison on an equal footing with the French compounds. In this type of engine the coupled wheels were placed close together, and by bringing the bogie as near as was possible to them in front, and placing the firebox immediately behind the second coupled axle, a convenient length of boiler was obtained.

Dimensionally the Great Central engines may be regarded as typical. They had 19 in. × 26 in. cylinders, 6 ft. 9 in. coupled wheels, boilers 15 ft. by 5 ft., and fireboxes 8 ft. 6 in. long,

23. Peter Drummond's *Castle* locomotive for the Highland. A 1913 version of a class introduced in 1900
24. North British Locomotive Company-built *Prince of Wales* class for the L. & N.W.R., designed by Bowen Cooke, introduced 1911

25. Robinson's mineral engine for the G.C.R., 1911
26 & 27. Pettigrew's passenger and goods engines for the Furness Railway, 1913

28. Bowen Cook's *Claughton*-class four-cylinder simple engine for the L.N.W.R., 1913
29. Midland designed goods engine for the Somerset and Dorset Joint Railway, 1914
30. Reid's passenger tank engine for the North British, 1915

31 & 32. Welsh and Scottish examples of the popular six-coupled radial tank

yielding a heating surface of 133 sq. ft. and a grate area of 26 sq. ft. Their total weight was 68 tons 5 cwt. with 37 tons over the coupled wheels. (Later engines of the class were heavier.) The firebox heating surface was somewhat small for an 'Atlantic', but this was because boilers of identical dimensions were fitted to the corresponding 4–6–0 engines on the same railway, which restricted its depth. The boiler of later 'Atlantics' built in 1906, when this type had been adopted in preference to the six-coupled engines for express work, had deeper fireboxes providing 153 sq. ft. heating surface.

The North Eastern and North British engines were larger. Both had cylinders 20 in. × 28 in., boilers 5 ft. 6 in. in diameter and fireboxes 9 ft. long, yielding in the former case 180 and in the latter 184·8 sq. ft. heating surface. The grate areas were 27 and 28·4 sq. ft. respectively. A further series of 'Atlantics' conforming to this plan was produced by Vincent Raven, Worsdell's successor, on the North Eastern in 1908.

Wide Firebox

A notable development of the 4–4–2 type of engine took place in 1902 with the appearance of Ivatt's first large 'Atlantic', in which the firebox was extended over the frames, the casing being only 5 ft. 11 in. long but no less than 6 ft. 9 in. wide. Apart from giving a substantially increased grate area—in this case 31 sq. ft.—the plan removed the restriction on boiler diameter imposed by the practice, to which Ivatt adhered till the last for all his engines, of inserting the inner firebox from the bottom—a constructional convenience which was only possible with the box between the frames where the barrel did not exceed some 4 ft. 10 in. in diameter, and from which other designers must have departed with reluctance. Ivatt took advantage of the removal of this restriction by providing his new engine, No. 251, with a boiler of 5 ft. 6 in. in diameter, the largest at the time of its appearance.

A minor disadvantage of the plan was that the firebox was shifted back, its width requiring it to be placed behind the wheels instead of immediately behind the axle, with a corresponding increase in the length of the tubes. Notwithstanding the very small clearance between the coupled wheels—the wheelbase was 6 ft. 10 in. and the wheels 6 ft. 7½ in. in diameter—and between the trailing bogie and leading coupled wheels, the distance between the tube-plates was 16 ft. The wide firebox design was, however, the only one which fully exploited the advantages of the 'Atlantic' type, and it is surprising that it was not more extensively adopted. The only designer to follow Ivatt was D. E. Marsh, who before his appointment at Brighton was principal assistant at Doncaster and whose 'Atlantics' of 1905 on the London, Brighton & South Coast Railway were direct derivatives from the Great Northern engines.

Multi-cylinder 'Atlantics'

As has been seen in the case of the Lancashire & Yorkshire engines, the combination of inside cylinders with the 'Atlantic' type, necessitating the use of the front coupled wheels as the drivers, produced an engine with a considerably longer wheelbase. This general layout was exemplified by the two very fine compound 'Atlantics', No. 730 and 731, produced by the North Eastern in 1906, in which the drive from all four cylinders was taken by the leading coupled axle. Their boilers differed considerably from those fitted to the corresponding simple engines. The diameter of the former was 5 ft. (6 in. less) but the grate area was 29 sq. ft. (2 sq. ft. more) and they were unique for the North Eastern in being provided with Belpaire fireboxes. (It may be added that, although compounding was not pursued further on this railway, front axle-driven 'Atlantics'—with three high-pressure cylinders—ultimately became standard.)

The lengthening of the wheelbase was virtually avoided (the bogie was set only 2 in. further forward) in the Great Western four cylinder simple of 1906, No. 40, *North Star*, in which divided drive was adopted and the inside cylinders were placed, not in their customary position over the bogie centre, but well ahead over the leading bogie axle. Basically, therefore, the layout was that of the two-cylinder 'Atlantics' on the same railway. So also was that of the four Great Central three-cylinder compounds of 1905–06 and the converted three-cylinder simple of 1909, the inside cylinder being placed under the smokebox and connected to the first pair of coupled wheels without any alteration of the wheelbase.

In every form of 'Atlantic' engine so far considered the back tube was placed behind the rear coupled axles. No. 1300, the Vulcan four-cylinder compound purchased in 1908 by the Great Northern for comparative tests with the 'Atlantics' of its own design, provided the solitary exception to this rule. In this case the tube-plate was brought forward, occupying a position between the coupled axles, with the result that the tubes were comparatively short (11 ft. 11 in.) and the firebox very long (10 ft.). The latter provided a grate area of 31 sq. ft., and 170 sq. ft. of heating surface—nearly 30 sq. ft. more than was provided by the wide fireboxes of the same company's standard 'Atlantics'. The boiler diameter was only 5 ft. 1⅝ in., but the firebox casing was considerably wider, the effect being similar to that obtained by the use of a centrally-coned boiler.

If the weakness of the 4–4–0 wheel arrangement was the restriction it imposed on firebox dimensions, that of the 'Atlantic' was lack of adhesion, which was inadequate for full use of the enhanced boiler capacity. The maximum weight allowed on each axle was in the neighbourhood of 20 tons, and in regard to the total weight available for adhesion the 'Atlantic' showed no advantage over the 4–4–0, while, the coupled wheels being closer, the position as regards ton per foot was worse.

4–6–0s

The Normal Pattern

The replacement of the trailing axle by a third coupled axle eliminated this defect, though it created other difficulties. The coupled wheelbase was considerably increased, as was also the internal resistance of the machine, while the presence of a coupled axle beneath the grate restricted the depth of the firebox. Nevertheless, the growing weight of the trains compelled many designers to have recourse to the six-coupled engine for the heaviest duties, and our period witnessed the introduction and increasing use of the 4–6–0 locomotive on many lines.

As was stated in chapter 3 this wheel arrangement was first adopted in Great Britain in 1894, on the Highland Railway by David Jones, who was followed by William Dean with his double-frame inside-cylindered No. 36 of 1896, and the much larger No. 2601 in 1899—both typical products of the Great Western during the formative period of its highly-individual locomotive tradition. But all these were intended for goods traffic.

At the opening of our period, only two railways, the North Eastern and the Highland employed 4–6–0s for express work. By the end of it, 4–6–0s for express work had been introduced on the Great Central, the Great Western, the London & North Western, the London & South Western, the Lancashire & Yorkshire and, in Scotland, on the Caledonian, and the Glasgow & South Western.

This type was used with triumphant success on the Great Western, but with varying degrees of success elsewhere. Nevertheless, by 1910 the perceptive observer would have regarded the

4–6–0 as the type destined to replace those with four-coupled wheels, whether of the 4–4–0 or 'Atlantic' order, for heavy-duty—which, indeed, in course of time, it did.

This wheel arrangement involved interesting problems (variously solved) of general planning and arrangement. That most commonly favoured included the outside cylinders driving the middle coupled axle, and the boiler with a firebox behind, but brought as near as possible to, the same axle. The first and second coupled wheels were placed close together, but the second and third were usually placed further apart to allow greater freedom in the design of the firebox, which sloped upwards from the front over the third axle.

The Great Central (Plate No. 19, depicting the earlier 'Fish' engines), Glasgow & South Western, and Great Western engines conformed to this pattern, which had already been adopted by the Highland and North Eastern, save that in the first and last named the coupled wheelbase was equally divided. The Great Central locomotive corresponded in every possible respect with the 'Atlantics', whose dimensions have already been given, but the 4–6–0s had 53 tons 7 cwt. available for adhesion and weighed 66 tons 16 cwt.

James Manson, on the Glasgow & South Western, contented himself with a 4 ft. 8 in. boiler and a grate area of 25 sq. ft. It may be noted that Peter Drummond's 'Castles' on the Highland had grates of 26·5 sq. ft., in this respect surpassing both the Great Central and Glasgow & South Western engines. A certain freedom in the design of the firebox was obtained by placing the second and third coupled axles 8 ft. 3 in. apart, this being facilitated by the small diameter of the wheels, 5 ft. 9 in., which enabled the first and second pair to be brought within 6 ft. of each other.

Like its somewhat ungainly predecessors of the same wheel arrangement, to which reference has already been made, William Dean's express 4–6–0 No. 100, of 1902, belongs to the formative period of Great Western locomotive history, although this machine is much nearer—indeed, all but exemplifies—the final product. Gone are the outside frames and inside cylinders which characterized the earlier machines; outside cylinders driving the central axle, inside frames, and a general layout of the kind now being considered are adopted.

Nevertheless there were features of marked originality. The cylinders were cast in pairs with half the smokebox saddle and bolted together at the centre line. The valves were outside. The main frames terminated just ahead of the leading coupled axle and had slab-like pieces bolted to them which passed under the smokebox portion of the cylinder castings, to which they were attached, and then curved upwards to meet the buffer beam. A piston stroke of 30 in. (already noted in the later 'Counties') was used for the first time. All these features became standard practice on the Great Western for two-cylinder 4–6–0s. The boiler barrel of No. 100 was not coned and the belpaire firebox casing towered above it. This was replaced in the second, and all subsequent, engines by a coned boiler and firebox with gently curving sides of the lateral dimensions adopted in the above-described 'Cities'. But the boilers for the 4–6–0s were longer, the barrel measuring 14 ft. 10 in. and the firebox 9 ft., giving a heating surface of 154·7 sq. ft. and a grate area of 27 sq. ft., the last two vital dimensions being greater than in any of the 4–6–0s so far noted.

Inside Cylinders

The problems which beset the designer of an inside-cylinder engine of the 4–6–0 type are well exemplified by the different manner in which they were solved (or left unsolved) in locomotives on this pattern which appeared during the Edwardian era on the Caledonian and the London & North Western Railways.

In the Caledonian design, the first of which, No. 49, appeared in 1903, the throat-plate was behind the second coupled axle, thus occupying the position adopted in the outside-cylinder 4–6–0s and the inside-cylinder Lancashire & Yorkshire 'Atlantics'. J. F. McIntosh, their designer (who is said to have favoured the 'Atlantic' type at first), seems to have been little concerned in avoiding the disadvantages inherent in the use of inside cylinders. The coupled wheels were equally spaced over a wheelbase of 15 ft. and the cylinders, placed centrally over the bogie, were 10 ft. 6 in. from the front coupled axle—a dimension typical of 4–4–0 engines, in which there was no need to restrict the length of the engine at this point. The result was a somewhat short connecting rod and an undesirably long wheelbase and boiler barrel; the wheelbase was 28 ft. 8 in. and the boiler 17 ft. 4½ in. An excessive length of tube was to some extent avoided by recessing the tube-plate into the barrel and so reducing the distance between the tube-plates to 16 ft. 3 in.

Subsequent engines of the same general design, built in 1906, were, however, given an extra 4 in. between the cylinder centres and the driving axle (permitting the use of 6 ft. 10 in. instead of 6 ft. 6 in. connecting rods), the wheelbase between the first and second coupled axles being correspondingly reduced. An important structural improvement was the lengthening of the driving wheel bearings—always a source of weakness in large inside-cylinder engines. This was rendered possible by the employment of smaller cylinders, with their centres placed closer together, and dishing the wheels outwards from circumference to centre. At the same time a larger boiler, 5 ft. 3½ in. instead of 5 ft. in diameter, was fitted, the length between tube-plates was increased by 5 in. and the firebox heating surface was raised from 145 to 148 sq. ft. These engines were beautifully finished and were the most imposing of the many fine products for which McIntosh was responsible.

George Whale in his 'Experiments' on the L.N.W.R.—the first of which appeared in 1905—tackled the problem very differently. A boiler of normal length—13 ft. between tube-plates—was combined with a firebox 8 ft. 2 in. long. The effect of this on an inside-cylinder engine, necessarily somewhat lengthy between the bogie and the leading coupled wheels, was to bring the throat-plate ahead of the middle coupled axle with consequent restriction on the depth of the firebox. The grate was, in fact, only 4 ft. 7½ in. below the centre line of the boiler.

The advantage of a normal length of barrel was thus purchased at the price of a shallow grate which, because of the position it occupied only a short distance forward of middle axle, could not be sloped down in front. (Such would have conflicted with Whale's invariable practice of providing horizontal grates, made, however, as deep as circumstances permitted.) These engines were of somewhat cramped design and their average performance fell below that of the 'Precursors', which is hardly surprising; for, while they had the advantage of some 4 sq. ft. in grate area, their firebox heating surface was nearly 30 sq. ft. less and their 6 ft. 3 in. driving wheels made greater demands on the boiler at similar speeds and cut-offs than the 6 ft. 9 in. wheels of the 'Precursors'.

Four Cylinders

The four-cylinder simple 4–6–0s which appeared during the reign of Edward VII—Churchward's brilliant 'Stars' on the Great Western, Hughes's massive products on the Lancashire & Yorkshire, with their 9 ft. 6 in. fireboxes yielding 190 sq. ft. of heating surface, and the two classes of Drummond engines on the London & South Western—may, as regards general layout, be readily assimilated to the designs in which only two cylinders were used.

The relationship of the 'Stars' to the two-cylinder 4–6–0s of the same railway was identical to that of *North Star* to the two-cylinder 'Atlantics'. In other words, for the reasons already given, they were basically of the outside-cylinder pattern. The L.S.W.R. engines, on the other hand, with their inside cylinders placed in the normal position below the smokebox and grates brought forward ahead of the central coupled axle, were basically of the form described in connection with the 'Experiments'.

The Lancashire & Yorkshire engines occupied an intermediate position. The inside cylinders, though not extending beyond the smokebox, were carried somewhat forward and the bogie was further from the coupled wheels than it need (and probably would) have been if only two cylinders had been used. In general, however, these approximated rather to the outside-cylinder than the inside-cylinder layout.

Locomotives of the same wheel arrangement with smaller coupled wheels suitable for fast goods, fish or excursion traffic, or for special road conditions, were to be found on the Caledonian, the Great Central and the London & North Western Railways, but these in point of design did not involve—or at least raise in so acute a form—the problems already discussed in relation to their express counterparts. Mention must also be made of Webb's 5 ft. 3 in. four-cylinder compound fast goods and excursion engine—representative of a superseded tradition and the last of that designer's products—which appeared in 1903 and in which the drive of all four cylinders was taken by the first coupled axle.

Goods Engines

Locomotives intended for heavy mineral and goods duties, and tank engines, which carried their supplies of coal and water on their own frames, can be dealt with more summarily, for in the former the problems of structure were less acute, because the lower axles enabled a firebox of tolerable depth to be placed in any relation thereto, while in both the usual practice was to incorporate a boiler corresponding to that of an express engine of similar capacity. In general, our period witnessed, as has been stated, an advance from six to eight coupled wheels for the heaviest mineral work, the classic British inside-cylinder 0–6–0—one of the most compact designs ever devised—being to that extent supplanted.

At the opening of the Edwardian era the 0–8–0 type had been used on the London & North Western for nine years, and on the Lancashire & Yorkshire for a short time, having been introduced on that railway in 1900. The North Eastern, Great Northern and Caledonian followed in 1901, the Great Central in 1902 (Plate No. 20) and the Hull & Barnsley in 1906. In the same year George Whale produced a large engine of this type by the rebuilding of one of his predecessor's three-cylinder compounds as a two-cylinder simple, and fitting it with a boiler of his standard diameter of 5 ft. 2 in. designed on similar lines to those used on his 4–4–0 and 4–6–0 locomotives. The boiler fitted to the 0–8–0 engines was longer than those of the other types but the firebox, as regards both length and depth was dimensionally intermediate. At the same time the Webb four-cylinder compounds were rebuilt with similar boilers, all being provided, before or on rebuilding, with a pair of carrying wheels to support the heavy cylinder castings in front. Four years later the first new 0–8–0 similar to the rebuilt three-cylinder engine was produced at Crewe.

The only other 2–8–0 class running during our period was that produced by Churchward in 1903 by the simple expedient of placing his 4–6–0 boiler over this arrangement of wheels—a matter of no difficulty, as the throat-plate fitted comfortably behind the third coupled axle, which took the drive from the outside cylinders of the standard pattern.

In the 0–8–0 the position of the cylinders had little or no influence on the general layout, for the firebox could as readily be placed behind the third axle and the grate raised over the fourth with outside cylinders driving the third axle as with inside cylinders driving the second. The former arrangement had the advantage of providing an ample length of connecting rod and both Wilson Worsdell on the North Eastern and Robinson on the Great Central, who had already resorted to outside cylinders on their larger bogie engines, not unnaturally favoured this plan. Ivatt, however, who was the first to adopt outside cylinders in the 'Atlantic' types, adhered to the inside position for his 0–8–0s, and he was the only designer to combine both positions in this way.

It is not surprising to find the inside-cylinder stalwarts—Aspinall, Whale and McIntosh, who refused to depart from this practice in their ten-wheeled engines—adhering to it for their 0–8–0s. The spacing of the wheels in the Caledonian engines is worthy of remark. The second and third axles were brought as close together as possible and the first and fourth placed at a considerable distance beyond them. This plan entailed a longer wheelbase but it gave more room for the machinery (apt to be cramped in inside-cylinder engines of this type) between the first group of axles, and enabled a deep firebox to be placed between the last. Inside cylinders were also provided on the Hull & Barnsley 0–8–0s, which had no passenger counterparts.

In 1907 George Hughes produced some four cylinder compounds on the Lancashire & Yorkshire with divided drive, thus combining both arrangements. Finally, the Great Eastern, which during our period had not advanced beyond the 4–4–0 for express work, but was to produce some inside-cylinder 4–6–0s shortly afterwards, presents a converse case to that of the Great Northern in adopting outside cylinders on its solitary 0–8–0, rebuilt from the 0–10–0 'Decapod' (to be described below); but this was doubtless due to the desire to retain as much as possible of the original engine.

Tank Engines

Several large tank engines were produced during the Edwardian period. Of these the 4–4–2 tanks on the London & North Western, the 0–8–2 tanks on the Great Northern and the 0–8–4 tanks on the Great Central were notable examples. The first two were but adaptations of the 4–4–0 and 0–8–0 tender engines on the same railways and involved no new problem of design. Nor did the Great Central engines, save for the provision of an additional cylinder between the frames (which did not involve any change in the basic layout of the corresponding 0–8–0) and the provision of an 'Atlantic' boiler, which was slightly larger than that of the 0–8–0s.

The Marsh 4–4–2 tanks on the London, Brighton & South Coast Railway presented no novel features structurally, being counterparts of the large Billinton 4–4–0s. One of them achieved fame from the remarkable economies obtained from the use of a superheater, demonstrated in comparative trials with a North Western 'Precursor'. Superheating, though the subject of experiments on a few railways during our era, belongs to the next chapter of locomotive history and calls for no elaboration here.

Once again the Great Western provides an exception to the general rule. The 2–6–2 tank on that railway—the first of which appeared before the coned boiler had become standard—was not developed from a corresponding 2–6–0; on the contrary the latter, shortly after the close of this period, was developed from the former and became the standard general utility engine on that line for many years.

Four other designs of tank engine produced during the period had no tender counterparts. The most remarkable was undoubtedly James Holden's 0–10–0 tank, which appeared in 1902 as a countermove to the projected electrification of the Great Eastern suburban lines. This engine had a wide firebox, after the pattern used in the large Great Northern 'Atlantics', and three cylinders, those outside driving the third axle, and that within the second by means of a forked connecting rod. This was the first and only time during the existence of the railways as independent companies that a wide firebox was placed over coupled wheels. The engine is said to have worked satisfactorily; but it proved too heavy for the road, and after a few years was reconstructed as described above.

The second design was H. A. Hoy's 2–6–2 tank on the Lancashire & Yorkshire, in which inside cylinders, driving the leading coupled axle, were combined with a boiler similar to that used on the 'Atlantics' but, having regard to the presence of coupled axles beneath it, with a somewhat shallower firebox. The third and fourth designs were the 4–6–2 and 4–8–0 tanks on the North Eastern, in each of which three high-pressure cylinders were employed. In the former the firebox occupied a position more or less central over a middle coupled axle—a plan adopted during the year succeeding our era by Robinson in his inside-cylinder 4–6–2 tanks on the Great Central and by S. D. Holden in his inside-cylinder 4–6–0s on the Great Eastern—a third variant of this pattern discussed above.

ENGLAND'S FIRST 'PACIFIC'

It is fitting that our review of Edwardian engines should conclude with a description of the most remarkable machine, in point of size and potential capacity, produced during the period. We have noted the tendency to replace the eight- by the ten-wheeled engine for express duty. In Churchward's *The Great Bear* of 1908 we find, for the first time in this country, matters carried a stage further in the production of a locomotive, carried on 12 wheels, which for those days was of phenomenal proportions and even today would be regarded as of great size. The 4–6–2 wheel arrangement secured at once the adhesive qualities of the 4–6–0 and the freedom to develop the firebox associated with the 'Atlantic'.

There was nothing novel, for the Great Western, about the machinery, which, save for an increase in the size of the cylinders, was similar to that of the 'Stars'. The adherence to this plan—horizontal cylinders over the bogie wheels—prevented an enlargement in cylinder volume commensurate with that in steam-raising capacity, and betrays a certain rigidity of design, or reluctance to depart from a plan which had proved so successful in the smaller engines.

The chief interest lies, of course, in the design and dimensions of the boiler. Here Churchward seems to have been at little pains to solve one problem to which the Pacific wheel arrangement gives rise. The wide firebox was necessarily placed behind the rear coupled wheels, the smokebox occupied its conventional position centrally over the bogie, and a large space was accordingly left to be bridged. In the result the boiler barrel was no less than 23 ft. long. The position could have been eased by the use of a combustion chamber and by setting back the smokebox or recessing the front tube-plate into the boiler. But none of these expedients was adopted. Here again we observe a certain want of flexibility, and it is difficult to avoid the impression that, had the same imaginative insight which marked the successful transition from the 'Cities' to the 'Stars' been brought to bear on the problems of the transition from the latter to *The Great Bear*, the distance between the firebox and the smokebox would not have been left to be bridged only by tubes of such excessive length.

The lateral dimensions of the boiler were very satisfactory and admirably demonstrated the freedom accorded to the designer by the 4–6–2 wheel arrangement. The barrel, composed of three rings, the central of which alone was coned, was 5 ft. 6 in. in diameter in front and 6 ft. at the rear. The firebox casing was 6 ft. 6 in. wide in front at the boiler centre and narrowed to 5 ft. 9 in. at the back, and was 8 ft. long. With a total wheelbase of 34 ft. 6 in., a total weight of

96 tons (60 tons being available for adhesion), a grate area of 41·79 sq. ft. and a total heating surface of 3,400 sq. ft. (including 545 sq. ft. furnished by the superheater and 158 sq. ft. by the firebox), this great engine far surpassed any running in this country at the time or destined to run for many years.

It would be pleasant to record that its success was proportionate to its size. But this was not so. Civil engineering restrictions confined it to the Bristol road and no opportunity was afforded in regular service for demonstration of its capabilities. Had *The Great Bear* been provided with fifty sisters and the necessary engineering works been undertaken to allow of their use on all the main routes of the Great Western system, there can be little doubt that, with certain modifications in design and after adequate testing at the Swindon plant and on the road, such a fleet would have opened a new chapter in the locomotive history of Great Britain. As it was, the engine ran for sixteen years, being refurbished and to some extent modified in 1913, and ultimately suffered the ignominious fate of being 'rebuilt' as a 4–6–0, which can have retained very little of the original engine.

So the two most remarkable Edwardian locomotives—Churchward's *The Great Bear* and Holden's 'Decapod'—were rendered comparatively ineffective by operating conditions. This, however, but emphasizes the limitations within which the designers of the period worked and reflects all the greater credit on them for the success they achieved in putting on the road locomotives capable of dealing effectively with the increasing weight of trains which was such a marked feature of the ten years of Edward VII's reign.

CHAPTER VI

The Last Years of the Companies

MORE than forty years have elapsed since the various railway companies of Great Britain lost their identity by being absorbed into one of the four groups constituted by the Railways Act, 1921, which came into operation on January 1, 1923—all save the Great Western, which with a number of Welsh lines, constituted a group in itself and retained the title under which it was incorporated in 1835.

The Great Western was one of the four largest systems. The other three, the Midland, the London & North Western and the North Eastern, were formed by the amalgamation of other lines, in 1844, 1846 and 1854 respectively, and were thus victims of the process to which they owed their existence. So also were the Great Eastern (1862), the Great Central (ex Manchester, Sheffield & Lincolnshire, 1849), the Lancashire & Yorkshire (1847, amalgamated however, with the North Western on the eve of grouping), the South Eastern & Chatham (worked by the Management Committee of 1899), the London, Brighton & South Coast (1846), the Glasgow & South Western (1850) and the Highland (1865). Other lines—like the Great Western—bore their titles from the beginning, attaining their full stature by new construction and absorptions. Such were the London & South Western (virtually, for the parent London & Southampton of 1838 assumed that title a year later), the Great Northern (1846), and the two largest Scottish lines, the Caledonian (1845) and the North British (1844).

All these and the lesser lines—the Great North of Scotland, the Furness, the North Staffordshire, the Somerset & Dorset Joint, the Midland & Great Northern Joint and the Midland & South Western Junction—to name only a few—possessed locomotives bearing unmistakable characteristics of ownership and in many cases of marked originality, which rendered the study of current practice one of absorbing interest. We are now concerned with principal developments which took place during the last twelve years of individual ownership by these companies. As before, emphasis will be laid on general features of design, and nothing in the shape of a complete chronicle will be attempted.

Superheating

The main feature which marked the opening of this period was the general adoption of the superheater. Preliminary work in this direction had already been carried out on some railways—notably the Lancashire & Yorkshire, the Great Western and the Brighton—and, before it opened, Earle Marsh's 4-4-2 T. on the last named was demonstrating in no uncertain fashion the economy and the increased power capable of being derived from the employment of this device.

Other railways quickly followed suit and by the end of the period a superheater had become a standard fitting.

The form ultimately adopted was the Schmidt—or direct derivatives therefrom—in which the superheater elements provided a double passage for the steam through the flues, so yielding a high degree of superheat. On the Great Western the flues housed three elements, each providing only a single passage for the steam, and the degree of superheat was lower. Preoccupation with compounding, as evidenced by the importation of the French 'Atlantics', may have led the Great Western to look askance at the emission of once highly superheated steam, still presumably capable of useful work, through the blast pipe; and moderate superheat, combined with the high working pressure employed on this line, gave very successful results, while lubrication problems with this combination were less acute; but it is difficult to resist the impression that in this case the majority view favouring a high degree of superheat was right.

Superheating was undoubtedly the greatest improvement effected in the steam locomotive during this period. Engines of the most diverse design were transformed by this device. The admirable 'Precursors' became the phenomenal 'George Vs'. The superheater version (Plate No. 24) of the underboilered, indifferent 'Experiments' were highly competent machines. At the other end of the scale the Great Northern 'Atlantics', almost too lavishly provided with the capacity to boil water, became the brilliant engines of later years. The story was the same on other railways, though the improvement was most obvious when the superheated engines had to be worked to capacity to keep time. But the increment in power, thus rendered so easily available, may well have had a hampering effect on constructional advances in the form of larger machines, such as was witnessed by the Edwardian era.

External Machinery

Perhaps the most significant development— though not the most obvious—which took place during the period was that productive of a form of locomotive in which inside machinery was totally eliminated. But readers should realize that the standard British plan pattern of locomotive with inside cylinders and motion continued to flourish (Plates Nos. 26–7, 30–2, 36). Our previous review showed how force of circumstances favoured the adoption of outside cylinders for the larger machines and how the Great Western went a stage further in removing the valves also to that position; and it is, perhaps, significant in view of what the future held in store, that the first to place the valve gear outside was no less a person than Nigel Gresley, then recently appointed Locomotive Engineer of the Great Northern. The pursuit of this line of development during the twelve years preceding the demise of the former railway companies will give an insight into current practice and may fittingly preface descriptions of other notable engines produced in the same epoch.

Apart from the feature to which attention has already been drawn, Gresley's 2–6–0 was a genuine Great Northern product, bearing all the marks of her Doncaster origin. The boiler diameter (4 ft. 8 in.) was restricted, as in all Ivatt's engines with fireboxes between the frames, to allow the inner firebox to be inserted at the foundation ring. Cab, chimney and dome followed the Ivatt pattern. The running plate was, however, raised to clear the coupled wheels and splashers were virtually eliminated—an admirable feature, which took full advantage of the accessibility secured by outside valves and gear. The wheel arrangement lent itself to the provision of a barrel of moderate, and a firebox of adequate, length (11 ft. 8 in. and 8 ft. respectively), the throat plate being brought as near as possible to the middle coupled axle, and the grate, of good depth in front, being raised by a gentle slope over the trailing coupled axle, which was

9 ft. distant from its companion. The grate area was 24·5 sq. ft. A high proportion of the total weight ($51\frac{3}{4}$ out of $61\frac{3}{4}$ tons) was available for adhesion, and the whole machine was admirably conceived to give efficient service on the mixed traffic duties for which it was intended.

A more powerful version, with boilers 5 ft. 6 in. in diameter (as on the same company's 'Atlantics') was produced two years later, being itself preceded by a 2–8–0 mineral engine with a 5 ft. 6 in. boiler, and cylinders, valves and gear similarly placed (Plate No. 33a).

Another early example of the 'all-outside' plan is furnished by R. W. Urie's 6 ft. mixed traffic 4–6–0 on the London & South Western, which came out in 1914. These engines, with their robust construction and large 21 in. × 28 in. cylinders, presented a marked contrast to their Drummond predecessors, all of which had four cylinders, divided drive, and fireboxes occupying a position partly over the second coupled axle, which restricted their depth at a vital point. The Urie engines followed approved practice in being provided with comparatively deep fireboxes with the throat plate behind the middle axle and the grate sloping somewhat sharply upwards over the trailing axle. A capacious firebox, 9 ft. long, furnishing 167 sq. ft. heating surface and 30 sq. ft. grate area, was thus provided, and the barrel (13 ft. 9 in.) was well proportioned in relation to it.

These good features were secured by a suitable disposition of the wheels, the front coupled axle being only 6 ft. 3 in. from the second, which was 7 ft. 6 in. from the third. The boiler dimensions were retained in the express version with 6 ft. 7 in. wheels, produced in 1918, but the extra space (9 in.) between the two leading coupled axles necessitated the cylinders being placed further forward and the provision of a longer smokebox, which was reduced in diameter by the coning of the first boiler course to save weight. A goods engine of the same general design but with 5 ft. 7 in. coupled wheels was produced in 1920, and the designer adopted the all outside plan for his 4–6–2 and 4–8–0 tank engines of the following year. In devising two standard boilers and two interchangeable sets of cylinders and motion for the five classes, and in providing them all with leading bogies and six-coupled wheels, this locomotive engineer showed a remarkable consistency in the excellent examples of advanced practice for which he was responsible.

Another railway to make exclusive use of this arrangement for new construction was the Highland, which also resembled the South Western in the invariable employment of a leading bogie. C. Cumming, the Locomotive, Carriage and Wagon Superintendent, produced a neat 4–6–0 goods engine and, shortly afterwards, an express counterpart in 1918. These followed the traditional plan with this type of engine where outside cylinders driving the middle coupled axle were employed. The throat plate occupied a position immediately behind the driving axle and the grate sloped upwards to clear the third.

These classes were preceded in 1916 by a remarkably interesting pair of 4–4–0 locomotives, providing, as they do, the first example (for main line work) of the all-outside plan being adopted without any inducement from considerations of general design, such as a desire to shorten the wheelbase by employing the second coupled axle as the driver. In the 4–4–0 engine the cylinders can as readily be placed between the frames as outside them, and while it is true that the location of the cylinders outside enabled the throat plate to be brought a few inches nearer the driving axle, since it was not necessary to set it back to clear a crank-axle, this was negligible in engines of such moderate dimensions as those under consideration.

The Cumming locomotives were not the first intended for the Highland to incorporate the all-outside plan, which had been adopted by G. F. Smith in his massive machines of 1915. But these proved to be too heavy for the Highland road and were sold to the Caledonian shortly after their construction.

The next engines of this group to be considered were of Midland design and represented a radical departure from the practice of that railway. In these locomotives their designer, Henry Fowler, united a superheated version of the boiler provided by Deeley on his 4–4–0 compound locomotive of 1905 with the 2–8–0 wheel arrangement for employment on the Somerset & Dorset Joint line (Plate No. 29). The boiler was an excellent steam-raiser, with its 9-ft. firebox, which was unsurpassed in British practice, and seldom equalled, in length at the time of its introduction, and provided 28·4 sq. ft. grate area. The adoption of boilers of identical dimensions on 4–4–0 and 2–8–0 locomotives was unique at the time, and remained so during the entire period of British railway history. (The converse case of the employment of the same boiler on 4–6–0 passenger and 0–6–0 goods engines will be noted in due course.)

Normally the eight-coupled goods engine was developed from an 'Atlantic' or 4–6–0 and carried the longer boiler associated with those types. A consequence of the adoption of the shorter boiler for the Somerset & Dorset engines was that the front of the firebox occupied a position ahead of the third coupled axle—not behind it as in all other cases—but, with the low axles resulting from the use of 4 ft. 7½ in. wheels, this did not restrict its depth, which was, of course, the same as on the corresponding passenger engine. The increased power rendered available by these locomotives must have been of the greatest possible assistance to the Somerset & Dorset, which heavy gradients notwithstanding, was otherwise operated by engines of quite moderate capacity.

The last engines calling for mention at this stage of our review were the products of R. E. L. Maunsell, which appeared on the South Eastern & Chatham in 1917. One of them was a 2–6–0 general utility locomotive with 5 ft. 6 in. wheels, the other a 2–6–4 tank, intended for passenger service, with 6-ft. wheels. With its coned barrel and Belpaire firebox sloping downwards and inwards towards the back, the boiler fitted to both classes was reminiscent of Great Western practice, though the ultimate Swindon refinements were not adopted, only the second boiler course being coned and the firebox plates being flat. It was, however, admirably proportioned and, for the classes of engine concerned, of good size. The boiler was 12 ft. 6 in. long and the diameters at front and back were, respectively, 4 ft. 7⅛ in. and 5 ft. 3 in. The 8-ft. firebox furnished 135 sq. ft. heating surface and a grate area of 25 sq. ft. Great Western practice was also followed in the provision of long-travel valves. Midland influence was evident in the smokebox door, which was flat and secured by six bolts on its periphery instead of by a central fastening and, it may be added, in the shape of the cab and tender sides. The 2–6–0 turned the scale at 59½ tons, of which 51 were available for adhesion. At the time of their appearance these engines were outstanding examples of advanced practice.

Deviations

The foregoing review shows that the 'all-outside' position was adopted for locomotives intended for all kinds of duty—express, goods and mixed-traffic alike. The development of the 2–6–0 for the last-named duty was a noteworthy characteristic of our period.

The pioneer in this field was the Great Western with its 'Moguls', which first appeared in 1911, and, as would be expected, exhibited the features associated with the locomotive practice of that company—coned boiler, long piston stroke, cylinders and valves outside, and valve gear between the frames. A boiler similar in all respects to that fitted to the 'Cities' and 'Counties' was dimensionally of perfect length for a 2–6–0 locomotive. Cylinders and valves of the design employed on the 2 cylinder 4–6–0s, the 'Counties' and the 2–8–0 mineral engines—cast in pairs with half the smokebox support and bolted together at the centre—were adopted, while the diameter of the coupled wheels, 5 ft. 8 in., was virtually midway between that used for the express and mineral engines. These 2–6–0s, proved to be a most useful class, capable of handling efficiently goods and passenger trains alike, and were a great acquisition to the admirable stud of locomotives possessed by this enterprising company.

A much larger mixed traffic engine, of the 2–8–0 type fitted with a boiler similar to that provided on the 4–6–0s, appeared in 1919. Two years later this engine was fitted with a larger boiler having a maximum diameter of 6 ft. and a firebox 10 ft. long, furnishing 170 sq. ft. heating surface and 30·28 sq. ft. grate area. So equipped the engine weighed 82 tons, of which nearly 73½ were available for adhesion. The distance between the cylinder centres and driving axle, 12 ft., was the same as on the 2–6–0s, and on a series of 2–8–0 tanks, carrying 'City' boilers, the first of which appeared in 1910, for the South Wales coal traffic. In each of these eight-coupled engines the second axle was the driver, in contrast with the 2–8–0 mineral engines, in which the third was so employed.

A second deviation from the 'all-outside' position—the converse of that employed on the Great Western—was to place the valve gear outside and the valves between the frames, as exemplified in L. B. Billinton's large 4–6–4 tanks for the Brighton. The explanation of this rather odd arrangement is probably derivative. The second of two similar, but smaller, 4–6–2 tanks, for which Earle Marsh was responsible, had the same arrangement, but the pioneer engine had the valve gear inside also, and, when the changeover was made in the second engine, it was probably found convenient to retain the cylinder castings. Robert Whitelegg's 4–6–4 tanks for the London, Tilbury & Southend had valves and gear between the frames, but on his later engines of the same wheel arrangement for the Glasgow & South Western both were outside (Plate No. 38). So also were they on the Metropolitan 4–4–4 tanks of 1920. In contrast the Furness 4–6–4 tanks of the same year had inside cylinders.

New designs during the period combining outside cylinders with inside valves and motion were the London, Brighton & South Coast 2–6–0s of 1913, the Caledonian '60' class 4–6–0s, introduced by William Pickersgill in 1916 and Robinson's 2–8–0 for the Great Central (Plate No. 25).

Inside Cylinders

Still more remarkable was the reversion to inside cylinders on the Great Central in its massive 'Sam Fays' of 1912–13. Some explanation of this *volte-face* may be furnished by the successful employment of inside cylinders with six-coupled wheels a few years earlier in the same company's 4–6–2 tank engines, although here the difficulties associated with inside cylinder 4–6–0 engines were not apparent, since the increased wheelbase could be utilized in providing a longer bunker. The 'Sam Fays', upon which for the first time a boiler 5 ft. 6 in. in diameter was employed by the Great Central, had fireboxes no longer, and grate areas no bigger, than those provided on their smaller predecessors, and no attempt was made to keep down the length of the barrel, which measured 17 ft. 3 in., the centre of the smokebox projecting some distance beyond the bogie pivot. If the term 'ungainly' could be used of any of Robinson's designs, this was it. Needless to say, these engines, like all those of the Great Central, were brightly and beautifully finished. A corresponding mixed traffic engine, with 5 ft. 7 in. coupled wheels, instead of 6 ft. 9 in., appeared shortly afterwards. It is of interest to recall that in his last 4–6–0 design—a mixed traffic engine produced in 1917—Robinson reverted to outside cylinders.

Another design of inside cylinder 4–6–0, which presented a marked contrast to that just considered, was that of S. D. Holden for the Great Eastern. Here the tension between wheelbase and boiler was resolved by placing the firebox almost midway over the second coupled axle. This gave ample room for the machinery—the wheelbase between the rear bogie and leading coupled axle was 8 ft.—and a reasonable length of boiler (12 ft. 6 in.), but resulted in the back of the firebox being a long way from the end of the frame, the provision of an enormous cab and lightly-loaded rear coupled axles, which carried only 14 tons. In fact only 44 of the 64 tons at which these engines turned the scale in working order were available for adhesion; but with their 20 × 28 in. cylinders, 26·5 sq. ft. grate area and 5 ft. 1⅛ in. diameter boilers, these 4–6–0s provided a notable increment of power compared with the 4–4–0 'Claud Hamiltons', and performed with marked success over the somewhat light road for which they were planned. An illustration of the compact design of the boiler is furnished by the fact that one of identical dimensions was provided for the company's largest 0–6–0 goods engines—in contrast with the Midland practice, which, as we have seen, provided the same boiler for a 4–4–0 and a 2–8–0.

The 4–4–0 with inside cylinders made a somewhat sensational re-entry on the Great Central with the very competent 'Directors' of 1913. Robinson's earlier 4–4–0s had given way in the Edwardian era to the 'Atlantics' and 4–6–0s introduced in 1904, the former being ultimately chosen for express work, which was not surprising with this company's lightly-loaded trains in those days. In reverting to the 4–4–0 type Robinson produced on eight wheels an engine of equal power to the 'Atlantics'—as a comparison of the boiler dimensions will show. The barrel, 12 ft. 3 in. long, was 2 ft. 9 in. shorter than that of the 'Atlantics', but had the advantage of 3 in. in diameter (5 ft. 3 in. against 5 ft.). In each case the firebox was 8 ft. 6 in. long and the grate area 26 sq. ft., the heating surface, 157 sq. ft., being 4 sq. ft. more in the 4–4–0s than in the later 'Atlantics', which was some 20 sq. ft. more than that provided by the original boiler fitted to both the ten-wheeled classes. The 4–4–0s, which weighed 61 tons (of which 39½ was available for adhesion), were some 10 tons lighter than the 'Atlantics,' and on the road proved to be fully their equals in performance. The same boiler was used on a 2–6–4 mineral tank engine with 5 ft. 1 in. wheels actuated by inside cylinders, the first of which appeared in 1914. The inside cylinder design also flourished in some large engines of the 4–4–0, 0–6–0 and 2–6–0 types produced by Peter Drummond for the Glasgow & South Western.

Four-Cylinder Engines

We now enter upon another chapter of locomotive development during this period. The tendency to employ more than two cylinders on locomotives intended for the heaviest duty, which had manifested itself during the Edwardian era, was accelerated during the last years of the individual companies, and several notable designs of three- and four-cylinder engines now demand attention.

The Great Central produced four-cylinder versions of its 'Sam Fay' (Plate No. 34) and corresponding mixed traffic classes (Plate No. 35). In these engines the four cylinders were placed in line, those between the frames driving the first, and those outside the second, coupled axle. Only two sets of motion, placed between the frames, were used for the four valves, the inside and outside valves on each side of the engine being attached to the upper ends of a Y-shaped pendulum, the lower end of which was attached to the Stephenson link motion, the necessary movement of the pistons in opposite directions being obtained by providing one valve with outside, and its companion with inside, admission.

Another noteworthy four-cylinder 4–6–0 class was the London & North Western 'Claughtons' (Plate No. 28). Here again the cylinders were placed in a line below the smokebox and only two sets of valve gear were employed, but in this case all four cylinders drove the leading coupled axle, and the valve gear was outside, communicating with the inside valve by means of rocking shafts. Since the front axle was the driver, the problems associated with the inside cylinder 4–6–0 plan presented themselves. In this case, however, in contrast with the same company's 'Experiments', the firebox design was not allowed to suffer. It lay behind the second coupled axle, and was consequently of ample depth, and extended well behind the third axle. It was 9 ft. 6 in. long, and provided a grate area of 30·5 sq. ft. and a heating surface of 171 sq. ft.—very generous dimensions for the period. Ample space was provided for the machinery, the wheelbase between the rear bogie and the leading coupled axle being 7 ft. 8 in. At the same time the boiler, 14 ft. 5⅞ in., was not inordinately long. These advantages were secured at the cost of setting back the smokebox from the conventional position (in contrast to the 'Sam Fays' already noted). The boiler diameter, of a maximum of 5 ft. 2 in., was the same as that employed on other North Western engines, but the Belpaire firebox was raised a few inches above it and was correspondingly wider at the sides, so providing in some degree the advantages of a coned barrel, without the constructional elaborations entailed with that form of boiler. It was, perhaps, fitting that the last express design emanating from Crewe should be the subject of acute controversy. When new and in capable hands, the 'Claughtons' were responsible for prodigies of haulage power and speed, probably unequalled in Great Britain at the time. Whether their subsequent falling off was due to lack of proper maintenance or to some inherent defect in design, the present writer is unable to say.

The Hughes four-cylinder 4–6–0s of 1921 were not new engines but rebuilds of those put on the road by the same designer thirteen years earlier. But the very substantial nature of the alterations and the fact that these locomotives were the prototype of new machines subsequently constructed would render their omission inappropriate. Piston valves replaced balanced slide valves and outside Walschaerts gear the Joy inside gear previously employed, and the cylinder diameter was enlarged from 16 to 16½ in. In these engines divided drive was employed with inside cylinders placed slightly ahead of those outside, but not so as to deprive the last-named of lateral support. The boiler was of moderate length (14 ft. 8 in. between tube plates) but the firebox was of large capacity—9 ft. 6 in. long and affording 176 sq. ft. heating surface. The coupled wheels, 6 ft. 3 in. in diameter, were disposed over the relatively short wheelbase of 13 ft. 7 in., with the consequence that the overhang at the trailing end, 7 ft. 2¾ in., was greater than usual. The boiler diameter, 5 ft. 9 in., was the same as on the original engines and the inside cylinder 0–8–0 and 0–8–2 tank, for which the same designer was responsible.

Boiler diameter, like firebox heating surface and grate area, is one of those criteria from which boiler capacity may be deduced (so far as dimensions can provide a clue), and it is significant that the standard adopted at Horwich exceeded that of any other British company. Comparative figures for other leading railways were: Great Central, Great Western, Great Northern, North Eastern and North British 5 ft. 6 in., London & South Western 5 ft. 4⅜ in., London & North Western 5 ft. 2 in., Great Eastern 5 ft. 1⅛ in. and Midland 4 ft. 9⅛ in.

Another arrangement of four-cylinder propulsion was employed in the unique 0–10–0 locomotive built at Derby in 1919 for banking trains up the Lickey incline on the Birmingham–Gloucester road of the Midland. Here again the cylinders were placed in line, but in this case they all drove the middle axle, and each pair was provided with only one valve, which was cross-ported for the inside cylinder. The cylinders themselves had to be raised considerably—the slope was 1 in 7—to prevent the inside connecting rod fouling the second axle, which, even so, had to be cranked to afford the necessary clearance. The boiler, which had a maximum diameter of 5 ft. 4 in., was the largest ever constructed at Derby till that time and the Belpaire firebox plates were sloped downwards at the top and inwards at the sides to enable a back plate of the dimensions employed on the same company's compounds to be used. It was 10 ft. long and provided a grate area of 31·5 sq. ft. The five coupled axles were disposed equally over a total wheelbase of 20 ft. 11 in. This engine was employed successfully for many years on the work for which it was designed and provides an outstanding example of the adoption of unusual means to meet a particular end.

A North Staffordshire 0–6–0 tank of 1922, in which the cranks on either side of the engine were set at 135 degrees, instead of the customary 180, so securing a very even turning movement at the cost of perfect balance and the necessity for four sets of valve gear, provides another solitary example of four-cylinder propulsion.

Three-Cylinder Propulsion

But it is the utilization of three cylinders on an extensive scale for engines planned for the highest power outputs that is the most notable feature of our period.

This method offered advantages over both the two-cylinder and four-cylinder systems, with quartered cranks, in providing a more even turning movement, and with it a more even draught on the fire. Moreover, compared with the four-cylinder machine, the increased cylinder capacity required was obtained at the cost of one additional cylinder, instead of two, while the three cylinders could be conveniently disposed within the lateral dimensions available under British clearances, and there was ample room for a strong crank axle and adequate main bearings. On the other hand, it lacked the (virtually) perfect balance of its four-cylinder rival, the outside cranks were necessarily set at 120 degrees and were thus less favourably placed to impart motion to the coupled wheels, and it required three sets of valve gear, or two and a conjugated gear, with the result that, reckoned by the number of working parts required, it offered no advantage over the four-cylinder system.

Some large 4–6–0s with this method of propulsion were introduced by William Pickersgill on the Caledonian in 1921—a line which, as we have seen, first favoured the inside and then the outside position for engines of this wheel arrangement. In the new engines the cylinders were arranged in a line below the smokebox, those outside, inclined at 1 in 69, driving the middle axle, while the inside cylinder, set at 1 in 41, drove the leading axle. This necessitated the bogie being placed well ahead of the leading coupled axle and, with the firebox behind the second axle, called for a comparatively long boiler—16 ft. between tube plates. The outside valves were actuated by Walschaerts gear and the inside valve by levers deriving their motion therefrom. With 9 ft. 3 in. fireboxes, furnishing 170 sq. ft. heating surface and 28 sq. ft. grate area, boilers 5 ft. 9 in. in diameter, 81 tons total weight (60 tons available for adhesion) these, the last, were the largest passenger locomotives ever built for service on the Caledonian.

It was, however, on the East Coast lines that the three-cylinder engine achieved its most notable stage of development. But the systems employed by Vincent Raven of the North Eastern and Nigel Gresley on the Great Northern were as different as they could well have been. The former adopted horizontal cylinders in line, three sets of Stephenson link motion, all inside, and front axle drive (save in the case of his 0–8–0 mineral engines, where such a system would have been impractical, and the second axle took the drive from cylinders sharply inclined). The employment of such a system on the 'Atlantic' engines, which operated the bulk of North Eastern expresses during our period, involved a lengthening of the wheelbase between the rear bogie and front coupled axles by 1 ft. 6 in. compared with the same company's two-cylinder engines, but the distance between the tube plates, 16 ft. 2⅝ in., remained the same, the smokebox being set

33a. Gresley's second design of 2–6–0 for the G.N.R.
33b. L.N.E. version of Gresley's third design 2–6–0 class

34 & 35. Robinson's largest express and mixed traffic engines for the Great Central, 1920–21
36. Ultimate development of the McIntosh 4–4–0 by Pickersgill, Caledonian Railway, 1922

37. L.N.E. version of Gresley's G.N. Pacific, introduced 1922
38. Whitelegg's Baltic tank, G. & S.W.R., 1922

39. Maunsell's 4–4–0 for the Southern, 1926
40. Fowler's L.M.S. *Royal Scot*, 1927

back on the newer machines. Raven adopted the same system on some 4–6–0 mixed traffic engines, on two classes of tank locomotives, of the 4–6–2 and 4–4–4 orders, and finally on the 'Pacific' express engines of 1922, to be noted presently. Only in the 4–8–0 tank engine was the drive divided, the outside cylinders actuating the second coupled axle.

On the Great Northern, progress took place in two stages. The first was the adaptation of the 2–8–0 mineral design to three-cylinder propulsion in 1918. All three cylinders were placed in line, inclined at 1 in 8, and drove the second coupled axle. To facilitate this arrangement this axle was placed 6 in. further back than on the two-cylinder engines, which had third axle drive, but the total wheelbase remained unaffected, since the second axle was now placed 6 in. nearer the third. As on all Gresley's three-cylinder engines only two sets of (outside) valve gear were employed, the inside valve deriving its movement from the outside valves. In all subsequent engines this movement was imparted by two levers attached to the outside valve spindles and to each other in such a way as to communicate the necessary movement to the inside valve.

The second stage was reached by dropping the outside cylinders to a more conventional—virtually horizontal—position and making a slight departure from the 120 degrees crank setting to allow for the incline of the inside cylinder, still, of course, necessitated by the presence of a coupled axle ahead of the driving axle.

This plan was adopted for the later mineral engines, and was a practical necessity on locomotives with larger wheels, such as the 2–6–0 of 1920 (Plate No. 33a) and the 'Pacific' of 1922 (Plate No. 37).

The former deserves more than a passing reference. Apart from three-cylinder propulsion, the design represents a substantial development of Gresley's original 'Moguls' of 1912 and the enlarged version of 1914. The first of these, it may be recalled, had boilers 4 ft. 8 in. in diameter. In the second this was enlarged to 5 ft. 6 in., but in the new engines it was 6 ft.—the largest ever employed on the Great Northern (and only once equalled elsewhere) for a barrel parallel throughout its length. The presence of a crank axle of course prevented the throat plate being brought as far forward as would otherwise have been possible, but this was mitigated by giving the throat plate a forward slope from the foundation ring, so that the back tube plate occupied in relation to its neighbouring axle a position similar to that achieved in locomotives with vertical throat plates adjacent to an uncranked coupled axle. The large diameter barrel was matched by a firebox of corresponding capacity, the latter providing 182 sq. ft. heating surface and 28 sq. ft. grate area. The coupled wheels carried 60 tons of a total weight of 71¾, and the design represented, in terms of power, the highest stage of development to which the 'Mogul' arrangement was brought, then or at any other time.

East Coast 'Pacifics'

Our review of the last years of the individual companies concludes, as did that of the Edwardian era, with an appraisement of locomotives of the 'Pacific' type. The Great Western engine was, as we have seen, abortive; but each of the two designs now to be considered would in all probability have become standard for first-line express work but for the inclusion of the two railways responsible for them in the same group, which entailed the elimination of one of them.

The North Eastern and Great Northern engines were of approximately equal size and capacity. With 19 in. × 26 in. cylinders, 6 ft. 8 in. coupled wheels and 200 lb. pressure, the former exerted a maximum tractive effort of 29,918 lb. The corresponding figure for the latter, derived from 20 in. × 26 in. cylinders, 6 ft. 8 in. coupled wheels and 180 lb. pressure, was 29,835. Each had 60 tons available for adhesion, the total weights being 97 and 92½ tons respectively.

The North Eastern boiler was 6 ft. in diameter throughout its length and housed 119 $2\frac{1}{4}$-in. and 24 $5\frac{1}{4}$-in. tubes. The distance between tube plates was 21 ft. and the grate area 41·5 sq. ft. The firebox furnished 200 sq. ft. heating surface, the tubes and flues 2,165 and the superheater 510—total 2,875.

The Great Northern engine was provided with a boiler varying in diameter from 5 ft. 9 in. in front to 6 ft. 5 in. at the back. There were 168 $2\frac{1}{4}$-in. and 32 $5\frac{1}{4}$-in. tubes 19 ft. long, and the grate area was 41·5 sq. ft. The total heating surface was 3,455 sq. ft., 215 being provided by the firebox, 2,715 by the tubes and flues and 525 by the superheater.

Both designers were at one in providing three high-pressure cylinders and wide fireboxes; but here all resemblance ceased. Raven employed the front axle as the driver, Gresley the second —a fact which is reflected in the longer wheelbase of the North Eastern engine—37 ft. 2 in. against 35 ft. 9 in. In the former the cylinders and valves, virtually horizontal, were cast in one, and three sets of Stephenson link motion—all between the frames—were employed. In the latter the inside cylinder was raised at 1 in 8 to enable the connecting rod to clear the leading coupled axle, and the valve, which lay beside it, was actuated by the derived motion already described.

The design of the boilers exhibited no less a contrast. That of the North Eastern engines was of the simplest possible form. A parallel barrel, 26 ft. long, was united with a round-topped firebox spreading out from the vertical by a few inches, and it was only by providing a substantial combustion chamber and recessing the front tube plate into the barrel that the tube length was brought down to 21 ft. The boiler barrel of the Great Northern engines was formed of two courses, the first parallel, the second coned about a central axis. The firebox sloped downwards and inwards from front to back, thus improving the outlook from the cab—a particularly valuable feature in a boiler of this size. But it was at the front end that the designer's ingenuity displayed itself to particular advantage. In the ordinary wide firebox, such as that fitted to the company's 'Atlantics', the throat plate sloped forward from the grate until it reached a point at which it joined the barrel, when it assumed the vertical. In the 'Pacific' engines the forward slope was continued to the centre line of the barrel, thus facilitating the provision of what was, in effect, a combustion chamber.

To sum up: One design exhibits a rigid—perhaps a too rigid—adherence to existing practice in the production of a new type; the other provides an outstanding example of the application of fresh thought to the new problem. The North Eastern 'Pacific' was an enlarged version of the same company's 'Atlantics' and 4–6–0s, the only new feature being the wide firebox. The retention of front-axle drive in particular created its own difficulties in a 'Pacific' engine by lengthening the wheelbase and increasing the distance to be bridged between the smokebox and firebox. The Great Northern engine, on the other hand, was planned afresh throughout. It is true that the cylinder arrangement had been used previously on the same company's 'Moguls', but it was one particularly well adapted to 'Pacific' requirements, since it allowed the middle coupled axle to be employed as the driver, with consequent shortening of wheelbase and tubes. Save as regards this feature, nothing like the new engine had been seen on the Great Northern before. The coning of the boiler about a central axis enlarged the cross-section of the combustion chamber, and this, with the elaborate shaping of the firebox above described, produced an excellent steam raiser, which, notwithstanding its great size, fitted neatly over the well-grouped wheels, and this engine provides an outstanding example of the solution of the basic problem in locomotive design—namely that of uniting the Stephenson boiler with wheels of sufficient size for fast work, all within the confines of the British loading gauge.

The reign of the North Eastern 'Pacifics' was a short one. The perpetuation of two such diverse designs within the group for identical duties would not have been justified. Gresley was placed in charge; and it is hardly surprising that his design was chosen. Grouping was productive of many such problems, and it would be rash to assert that the right course was invariably taken. But it is difficult to resist the conclusion that in this case the better design survived.

Great Northern, as Gresley's pioneer engine was appropriately named, proved to be the first of one of the most famous classes of all time. Suitably modified to cope with the more exacting duties laid upon them as time went on, these engines were to remain in the front rank for nearly 40 years, and, if they have at last yielded pride of place, it is not to one of their own kind, but to a form of motive power which owes nothing to the genius of Stephenson or to that body of men—highly competent, occasionally brilliant, well known or working in obscurity—who in their several ways adapted the crude prototype to the needs of their times and finally brought it to near perfection.

CHAPTER VII

Grouping

INTRODUCTION

THE quarter of a century, from 1923 to 1948, during which the railways of this country were operated as four groups, constituted by the earlier companies witnessed a number of important advances in locomotive practice.

No two groups pursued the same policy, though basic problems confronting three of them were the same. The Great Western, as the only large railway within its group, pursued under C. B. Collett, who had succeeded Churchward in 1922, its own locomotive policy unchanged and, indeed, applied its characteristic features to many locomotives of the absorbed companies.

With the other three groups the conflict between the locomotive policies of the main constituents had to be resolved. The L.M.S., which inherited more than 5,000 locomotives from the North Western (including some 1,650 of Lancashire & Yorkshire origin), and more than 3,000, 1,000 and 500 from the Midland, Caledonian and Glasgow & South Western respectively (none of the other constituents providing as many as 200), pursued a vigorous policy of standardization combined with ruthless scrapping of much excellent material. Horwich had, surprisingly, gained an ascendancy over Crewe on the amalgamation of the Lancashire & Yorkshire and London & North Western railways and George Hughes was placed in charge of the group, with Henry Fowler of Derby second in command. But the Lancashire & Yorkshire influence on the locomotive policy of the L.M.S. was short-lived. Fowler succeeded Hughes in 1925 and his period witnessed the multiplication of Midland types, while the new designs were stamped with the hallmark of that railway's locomotive practice. The appointment of William Stanier from Swindon, in 1932, led to the introduction of a number of Great Western features and the production of a series of designs as attractive as any which had marked a locomotive chieftainship before, during or subsequent to our period.

The L.N.E. was provided with some 7,000 locomotives by its constituents (about 3,000 less than the L.M.S.). The North Eastern furnished more than 2,100, the Great Northern, Great Central and Great Eastern between 1,300 and 1,400 each and the North British nearly 1,000. The locomotive policy of the group was determined by the fact that Nigel Gresley, of the Great Northern, was placed in charge, but this engineer evinced a real appreciation of the worth of much of the material furnished by companies other than his own, particularly the Great Central, and new construction was not confined to locomotives of his own design. Nor did he insist on the details of his own designs conforming with Doncaster practice, the 2–6–0s built at Darlington exhibiting, for example, a number of North Eastern features.

The three constituents of the Southern group made a more even contribution to that railway's locomotive stock than did those of the northern lines. On the eve of grouping the South Western possessed (in round figures) some 900 engines, the South Eastern 700 and the Brighton 600. R. E. L. Maunsell, of the South Eastern, who was placed in charge, did not, perhaps unfortunately, pursue the lines laid down in his admirable 'Moguls' when more powerful machines were required, but found the prototype for his first express engine in another of the constituents with the result that locomotives of very different design were multiplied simultaneously and the Southern did not exhibit that consistency in locomotive practice which marked the other groups. His retirement in 1937 and the arrival of Bulleid was followed by the production of locomotives of quite startling originality.

Developments in practice during this quarter of a century will be most conveniently noted under the groups themselves.

I THE GREAT WESTERN

The principal feature of Great Western development was the exploitation of the 4–6–0 type. At the outset of our period that company possessed only two main classes, each of which carried the same boiler and differed only in the number of cylinders provided, four or two, the first being generally reserved for the heaviest work. By the end of the period six further classes had been introduced, the Churchward boiler surviving on only two of them (the 'Halls' and the 'Granges', which were, respectively, versions of Churchward's 6 ft. 8 in. two-cylinder engines with 6 ft. and 5 ft. 8 in. coupled wheels), while the boilers of the other four differed dimensionally *inter se*, though all retained the essentials of Churchward practice.

'Castles' and 'Kings'

The development of Churchward's four-cylinder 'Stars' took place in two stages. The first was in the production of the 'Castles' (Plate No. 45) in 1923, to which it had originally been intended to fit a boiler standard with that fitted on the mixed traffic 2–8–0s, but this was vetoed by the engineering department, and a boiler of intermediate size had to be provided. The original wheelbase was retained but the frame was lengthened to accommodate a longer firebox. The boiler diameter was increased from 4 ft. 10 $\frac{13}{16}$ in. and 5 ft. 6 in. at front and back to 5 ft. 1 $\frac{15}{16}$ in. and 5 ft. 9 in. The length remained the same (14 ft. 10 in.), so also did the number of the superheater flues (14), the extra diameter being utilized to provide 201 small tubes instead of 176. The firebox provided on the 'Castles' was a foot longer than on the 'Stars', the 10 ft. casing on the former providing a grate area of 29·36 sq. ft. and 163·7 sq. ft. heating surface, an increase of 2·6 and 9·4 sq. ft. respectively in these vital dimensions. These changes involved an increase of weight in working order of 4¼ tons, to 79 tons 17 cwt. With an enlargement of the cylinders to 16 in. × 26 in. the tractive effort at 85 per cent. of the 225 lb. per sq. in. boiler pressure was 31,625 lb., the highest figure then achieved on any express locomotive in the United Kingdom and the occasion of some ill-advised publicity.

Development was carried a stage further in the 'Kings' of 1927 (Plate No. 46), in which the Churchward basic dimensions were abandoned, and a boiler of unprecedented capacity for this type of locomotive was placed over a wheelbase of equally unprecedented length. Eight feet separated the first coupled axle from the second and 8 ft. 3 in. the second from the third. The bogie wheelbase was 7 ft. 8 in., and the total wheelbase, 29 ft. 5 in., was 2 ft. 2 in. longer than that of the 'Castles' and 'Stars'. The retention of horizontal cylinders (over the leading and trailing bogie axle) and their enlargement to 16¼ in. diameter rendered it necessary to place the bogie frame outside for the leading and inside for the trailing wheels—a unique feature.

But it is for their boiler dimensions that these engines are particularly notable. The firebox, outside, was no less than 11 ft. 6 in. long and furnished a grate area of 34·3 sq. ft., both of these being the largest ever employed in this country with a narrow firebox. The barrel was 5 ft. 6¼ in. and 6 ft. in diameter (another record for a 4–6–0) and 16 ft. long. The pressure was raised to 250 lb. per sq. in., which, with 6 ft. 6 in. coupled wheels, the lengthening of the piston stroke to 28 in. and the other relevant dimensions already mentioned, gave a tractive effort of 40,300 lb.—a figure never exceeded, and only once equalled, on an express engine in British practice. A moderate degree of superheat was provided by the 16 flues—two more than in the earlier 4–6–0 locomotives on this line—and the engine turned the scale at 89 tons, no less than 22½ tons resting on each coupled axle.

The 'Manors'

Collett's third new design of 4–6–0 (we do not include the 'Halls' and 'Granges' as such) was at the other end of the scale. The 'Manors' of 1938 were small engines weighing 68 tons 18 cwt. in working order. The wheels were spaced as on the larger two-cylinder 4–6–0s but the over-hang at the back was reduced from 6 ft. 6 in. to 5 ft. 3 in. The boiler was 12 ft. 6 in. long and tapered to a maximum diameter of 5 ft. 3 in. The firebox measured 8 ft. 8½ in. at the boiler centre line and the grate area was 22·1 sq. ft. The discerning will have noticed something odd about these dimensions. A firebox of this length would be expected to provide a grate area of 26–27 sq. ft. The boiler was shorter than would be expected for an engine of this type with a normal wheelbase. The explanation lies in the fact that the throat plate was given a forward slope between the foundation ring and the point where it met the barrel. This feature, which we have observed on the Gresley 'Moguls'—where the presence of an inside crank was an additional motive for its employment—improves the longitudinal proportions between barrel and firebox, without bringing the front of the grate over the middle coupled axle, a course which necessitates the employment of an unduly shallow firebox.

Hawksworth's 'Counties'

This plan was perpetuated by F. W. Hawksworth, Collett's successor, in the largest two-cylinder 4–6–0 ever produced by the Great Western, in 1945. These engines were provided with a firebox which furnished 169 sq. ft. heating surface and a grate area of 28·8 sq. ft. The maximum boiler diameter of 5 ft. 8⅝ in. is not, so to speak, a Great Western dimension, but Swindon had been concerned with the production of a number of 2–8–0s of Stanier design for war purposes with an identical boiler diameter, and it is not unfitting that this most eminent of Swindon's sons should, in this important particular, determine the dimension of the last design of 4–6–0 produced by the Great Western. Nevertheless, where Swindon and Stanier principles conflicted, the former prevailed; domes were not employed, and the 'Counties', as the new engines were named, were in the genuine Great Western line of development. Indeed, the Great Western characteristic of high boiler pressure was carried a stage further with the provision of a boiler pressure of 280 lb. per sq. in. However, with 21 flues a comparatively high degree of superheat was provided, in contrast with normal Great Western practice.

The most important constructional difference between the 'Counties' and their Churchward predecessors was the abolition of that designer's front end with the cylinders cast in pairs with half the smokebox saddle, bolted together at the centre, and auxiliary frames between the buffer beam and the main frames just ahead of the leading coupled wheels. These auxiliary frames necessarily lacked depth and called for reinforcement by bars, which were a characteristic, not to say a somewhat crude feature of the Great Western mixed traffic and mineral engines. In the 'Counties', as in the later 'Halls' and 2–8–0s built under Hawksworth's superintendency, the main frames were carried through to the buffer beam with the cylinders bolted to them in accordance with normal British practice. With cylinders 18½ in. by 30 in., coupled wheels 6 ft. 3 in. in diameter and the above-mentioned boiler pressure, the tractive effort was 32,500 lb. The total weight was 76 tons 17 cwt., of which 59 tons 2 cwt. was available for adhesion. The 'Castle' and 'Star' wheel-spacing was retained save for an extra two inches in the bogie wheelbase and a corresponding reduction of the measurement between the rear bogie and leading coupled axle.

Other Developments

It is not necessary to linger over other Great Western developments during our period. It will be sufficient to recall that the 0–6–2 tanks of 1924 and 0–6–0s of 1930 exhibited a new combination of coned boiler and inside cylinders; that the 0–6–0 pannier tanks of 1934 provided the first example on the Great Western of the employment of outside, and complete suppression of inside, machinery (cylinders, valves and gear), long after this had become a familiar feature on the other groups; the adaptation for main line work of the 2–8–0 tanks of 1910—no longer required for the South Wales coal traffic—by the provision of a trailing axle with an enlarged bunker carrying six tons of coal and the increase of water capacity from 1,900 to 2,500 gal.; the rebuilding of two series of 2–6–2 tank locomotives with smaller wheels to give enhanced tractive effort; and the provision of a more ample cab (of the 'Castle' pattern) on the 2–8–0 mineral engines, which, save for this and the constructional changes at the front end already noted, remained unaltered.

It is difficult, looking back over this final period of Great Western history, with all the advantages of one so situated, not to regret that the Churchward standards were not more rigidly adhered to. Apart from the 'Kings', which demanded a departure from the earlier standards, no Great Western locomotive produced in this period offered any real advantage over designs capable of being formulated within the lines already laid down. The large boiler fitted to the 2–8–0 mixed traffic locomotive could readily have been accommodated over a 'Star' chassis; the 0–6–2 tank and 0–6–0 goods engine could have offered few, if any, advantages over the Churchward 2–6–2 tanks or the excellent Dean 0–6–0s; the modification of the 2–6–2 tanks to provide extra tractive effort can be regarded—and perhaps only regarded—as fiddling; and it is doubtful whether the 'Manors' filled a need which could not have been as well met by the Churchward 2–6–0s, either of the smaller or larger boilered variety.

Nevertheless, all the new designs were of genuine Churchward vintage, and it is satisfactory to recall that the traditions of that great engineer initiated at the turn of the century remained with the Great Western to the last.

II LONDON, MIDLAND AND SCOTTISH

Hughes

Hughes's influence on London, Midland & Scottish Railway locomotive practice was evanescent. Some new 4–6–0s, corresponding generally with the rebuilt machines, became for a time the front-line express engines of the system and some large 'Baltic' tanks were developed from this design. But neither was subsequently multiplied or became an L.M.S. standard type.

It was otherwise with the 'Moguls', the first of which emerged from Horwich after Fowler, Hughes's deputy for the first two years, had taken charge, after a short interregnum by Ernest Lemon. These engines bore unmistakable signs of their place of origin in characteristics differing widely from Midland practice, which underlay all Fowler's other work. Boiler fixtures, cab and motion were of genuine Lancashire & Yorkshire vintage, though the design itself was entirely new, while the practised eye would discern Midland influence here and there, and the tenders were pure Derby.

The engines were planned on generous lines. The wheels were well spaced, by 9 ft., 8 ft., and 8 ft. 6 in. reading from front to rear, while the boiler had a maximum diameter of 5 ft. 5 in. and was provided with a firebox 9 ft. long, furnishing 27·5 sq. ft. grate area and 160 sq. ft. heating surface. The grate was of good depth—5 ft. 9 in. below the boiler centre in front and 4 ft. 6 in. at the back. The cylinders, 21 in. × 26 in., were considerably larger than any previously used with this type of engine and their large diameter necessitated their being steeply inclined in order to afford the required clearance.

The machinery was entirely external and the running plate was raised to clear it, assuming over the cylinders a height unprecedented in this country, which gave these engines a somewhat ungainly appearance and earned them the nickname of 'Crabs'. Nearly 250 engines of this class were built at Horwich and Crewe (Nos. 13000–244) before the design was modified according to the ideas of a succeeding locomotive chief, in 1933.

Fowler

Henry Fowler's work for the L.M.S. during his five years of office comprised the multiplication of three Midland types derived or developed from Deeley—the class '2P '4–4–0, '4F' 0–6–0, and, in a slightly modified form, the compounds (all equipped with superheater)—and in the production of five new designs.

The first two Midland classes carried boilers of identical dimension, 10 ft. 6 in. long in the barrel and 4 ft. 9½ in. maximum diameter, a 7-ft. firebox giving 21·1 sq. ft. grate area and 125 sq. ft. heating surface, and 21-element superheaters. The compounds had boilers of similar lateral dimensions, but they were some 1 ft. 6 in. longer, and the fireboxes measured 9 ft., with the result that the grate area was increased to 28·4 sq. ft. and the heating surface to 152·8 sq. ft. (Plate No. 22).

These sectional dimensions were retained by Fowler in his first design for the group—a 2–6–4 passenger tank engine—but the barrel was of the same length as those of the 2P and 4F class (which led to the chimney being set back from the cylinder centre line), while the firebox corresponded with that fitted to the Midland '700' class 4–4–0, which was of intermediate capacity between the '1Ps' and the compounds and had 25 sq. ft. grate area. The first four axles were spaced as on the 'Moguls' and the motion was similar. These engines differed from anything previously seen on the Midland, as the machinery was entirely external (though Derby had used this plan some years earlier in the 2–8–0s built for the Somerset & Dorset Joint Railway). They proved to be excellent engines in service with a remarkable turn of speed, having regard to their 5 ft. 9 in. coupled wheels (3 in. larger than those of the 2–6–0s). A smaller version, produced some three years later, in the form of a 2–6–2 tank locomotive calls for no further mention.

The mechanism of the 2–6–4 tanks and the Horwich 'Moguls', slightly modified, was adopted in one of the most remarkable classes of locomotives ever to run on British metals. Two of them, with cylinders reduced to 18½ in. diameter, and set almost horizontally in consequence, and with 5 ft. 3 in. coupled wheels, were set back to back and united by a girder frame affording support for a large boiler dimensionally unhampered by the presence of wheels beneath it.

The firebox, straight-sided and unwaisted, providing a grate area of 44·5 sq. ft. and 183 sq. ft. heating surface, was united with a barrel 6 ft. 3 in. in diameter and 12 ft. long, containing 36 superheater flues and 202 small tubes—a designer's ideal in simplicity of construction, overall proportions and effectiveness as a steam raiser. Thirty-three of these 2–6–0+0–6–2 Beyer Garratt engines were built by Beyer, Peacock, which specialized in this design, for the Nottingham–London coal trains and are said to have replaced double the number of 0–6–0s previously used on these duties.

A reversion to earlier practices was manifested in the next Fowler design to be noted. The North Western 0–8–0 mineral engines were wearing out and, when replacements were needed, recourse was had to the same wheel arrangement combined with inside cylinders. These were, in effect, a Midland version of the North Western 'G4s'. As before, the grate was horizontal and of an area of 23·6 sq. ft., but a Belpaire firebox was provided, together with a smokebox of typical Derby form. The pressure was raised by 20 lb. per sq. in. to 200 and, with the alteration in the size of the cylinder from 20½ in. × 24 in. to 19 in. × 26 in., did not differ greatly in nominal tractive power from their predecessors. The lengthening of the wheelbase by one foot between the first and second coupled axles gave much needed extra room for the machinery, particularly in view of the longer stroke, and Walschaerts gear was substituted for the Joy motion previously adopted.

These engines are of interest as providing the last design of the 0–8–0 inside-cylinder mineral engine in this country—a type which had been introduced by Webb three and a half decades earlier, and, if abandoned by him in favour of compounds, was extensively used by his successor on the North Western and on the Great Northern, Caledonian, Lancashire & Yorkshire and Hull & Barnsley railways.

The *Royal Scot* (Plate No. 40) was Sir Henry Fowler's *magnum opus*, though he would have produced a far larger machine if he had been given a free hand. However, the fifty engines of this class—all built by the North British Locomotive Co. Ltd. at short notice—were substantially more powerful than any other front-line express engine on the L.M.S. The 'Claughtons' were long past their prime and the Horwich 4–6–0s had not been a conspicuous success. Nor was the position improved by the multiplication of Midland compounds which—admirable engines as they were—had to be used in pairs on the heaviest expresses.

In the new engine Fowler provided three high-pressure cylinders with divided drive (differing from the compounds in both respects) and a boiler of near maximum capacity for a 4–6–0 type of locomotive. The diameter, 5 ft. 9 in., was virtually one foot longer than the Midland standard. The throat plate was given a forward slope, in the manner already described in connection with the Great Western 'Manors', and kept the length between the tube plates down to 14 ft. 6 in. The grate was of good depth, 5 ft. 8 in. below the boiler centre in front, and sloped up over the trailing coupled axle which, placed 8 ft. from the adjacent axle, provided ample room for a long firebox of 10 ft. 3 in. at the foundation ring, yielding 189 sq. ft. heating surface and 31·2 sq. ft. grate area. A high working pressure of 250 lb. per sq. in. was combined with a high degree of superheat furnished by 27 elements.

The smokebox was, in accordance with Midland practice, of considerably larger section than the barrel. The diameter was 6 ft. 7½ in. and as it was pitched 9 ft. 3½ in. from the rail level the chimney was diminutive, only a little over 7 in. Even so, the maximum height exceeded that usually allowed for locomotives intended to operate over the Scottish loading gauge, which ordinarily prescribed 12 ft. 11 in. as the limit. These engines weighed all but 85 tons.

Stanier

Stanier's first main-line engines for the L.M.S. were of the 'Pacific' type. But it will be convenient to open our consideration of his work with the 'Moguls' which followed them only a few months later. These engines presented a remarkable contrast to those which had preceded them and important, if less readily discernible, differences from those which followed. The contrast was the more marked because the new engines constituted a new series of 2–6–0s, closely resembling their forerunners in power and general arrangement. The wheel spacing was the same, so was the outside position of cylinders and valve gear, the Hughes cab was perpetuated, and the tenders were identical.

Swindon influence—in marked contrast to anything which had appeared on the L.M.S.—was evident in the design of the boiler, with its coned barrel and the radially curved sides and top of the Belpaire firebox, the partly horizontal and partly sloping grate, the high working pressure and low degree of superheat. But the boiler was not a complete reproduction of the Churchward pattern: the back plate sloped outwards from top to bottom, the firebox was not built out from the rear barrel plate at the centre line and the disposition of the boiler fixtures was different, for while the boiler feed pipes were led to the top of the barrel the safety valves were not combined with them but were placed in the customary position over the firebox. Great Western practice was followed in the horizontal disposition of the cylinders but the adoption of outside valve gear represented an important departure from Swindon practice, as did also the use of continuous frames from buffer beam to drag box.

Though the main characteristics of this prototype were to appear in all Stanier's standard engines, the designer himself was to make certain important changes in his later engines. The horizontal position of the cylinders was abandoned (doubtless because of the less generous L.M.S. loading gauge), and in course of time a high degree of superheat was provided and domes were fitted.

Dimensionally these engines were of considerable size and in power output did not differ materially from the 'Moguls' which they followed. Cylinders 18 in. × 28 in., coupled wheels 5 ft. 6 in. in diameter, and 225 lb. per sq. in. boiler pressure gave a tractive effort of 26,288 lb., compared with 26,580 for their predecessors, derived from 21 in. × 26 in. cylinders, 5 ft. 9 in. coupled wheels and 180 lb. per sq. in boiler pressure. The superheating surface dropped from 307 to 193 sq. ft., the grate area was 27·8 sq. ft. (fractionally more) and the weight 65 tons—a ton less.

The boiler barrel tapered from 5 ft. in front to 5 ft. 8½ in. at the back. Similar dimensions (the published figures differ fractionally) were adopted by Stanier for his two classes of 4–6–0—the three-cylinder 'Jubilees' and the two-cylinder general utility class '5'—and his 2–8–0 freight engine. The length between tube plates was 12 ft. 3 in. in the case of the 2–6–0 engines, 13 ft. 3 in. in the case of the 2–8–0s and 14 ft. 3 in. in the case of the 4–6–0s. All had 9 ft. 3 in. fireboxes giving 27·5 sq. ft. grate area and 155 sq. ft. heating surface, except the 'Jubilees', the fireboxes of which were 9 in. longer, the grate area and heating surface being correspondingly increased to 29·5 and 162·4 sq. ft.

In the modifications to Stanier's later 'Jubilee' and 'Black Five' 4–6–0s and the 2–8–0 freight engines, domes were provided and a forward slope was given to the throat plate from the foundation ring to the point where it met the barrel, and the tube length of the 4–6–0s was correspondingly reduced by one foot, bringing it into line with the 2–8–0s. In the last named, and in the class '5', the grate area was increased to 28·65 and the heating surface to 171 sq. ft. In the 'Jubilees' the corresponding figures were 31 and 181. The superheating surface was substantially augmented by the provision of 24 in place of the original 14 flues, and in the case of the latest examples of class '5', this number was further increased to 28.

Ultimately a large boiler was fitted to some of the 'Jubilees' and Fowler's 'Royal Scots'. Here the maximum diameter was increased to 5 ft. 10½ in., the length between tube plates was three inches less than in the former 'Jubilee' boiler and that of the firebox three inches more—13 ft. and 10 ft. 3 in. respectively. Firebox heating surface and grate area became 195 and 31·25 sq. ft. and 28-element superheaters were provided. The boiler pressure was increased to 250 lb. per sq. in. It is not surprising that with these dimensions (to say nothing of their excellent design) the engines, whether originating as 'Royal Scots' or 'Jubilees', proved to be one of the most redoubtable series of 4–6–0 locomotives ever to run in this country.

A boiler of similar design but considerably smaller, with a maximum diameter of 5 ft. 3 in., a firebox 8 ft. long and a grate area of 25 sq. ft., was provided for the Stanier version, both of the two-cylinder and three-cylinder variety, of Fowler's 2–6–4 tank locomotives and one smaller still—the corresponding dimensions being 4 ft. 9 in., 5 ft. 11 in. and 17·5 sq. ft.—for the 2–6–2 tanks.

Stanier used two, three and four cylinders over the range of his designs and has sometimes been regarded as inconsistent in so doing. But the reasons for this variation are not difficult to discover. With one exception he adhered to the two-cylinder plan (with the 'all outside' position) where sufficient power could be obtained with moderate cylinder diameter. For the larger engines he adopted one or two extra cylinders in accordance with the power required. The only exceptions to this rule were the three-cylinder 2–6–4 tanks (Plate No. 42), and the probable explanation here is the desire to utilize to the full the weight available for adhesion—a particularly valuable feature for locomotives planned to haul frequently-stopping passenger trains. It may be noted in passing that their two-cylinder counterparts had the largest cylinder diameter ever used by this designer: 19 in.; 17½ in. sufficed for the 2–6–2 tanks, 18 in. for the 2–6–0s, 18½ in. for the 2–8–0s and class '5s', 16–17 in. for the three-cylinder 4–6–0s and 16¼–16½ in. for the 'Pacifics'. The 2–6–2 and 2–6–4 tanks and the three-cylinder 4–6–0s had 26 in. stroke, the rest 28 in.

The 'Pacifics'

Our final chapter in Stanier's locomotive career concerns the introduction and development of the 'Pacific' type for the most exacting express work; and here again we shall observe the extensive adoption of Swindon practice, followed in some degree, and with one qualification, by its abandonment.

Indeed, in the first engine which appeared in 1933, preceding the 'Moguls' by a few months, the mechanical details and wheel spacing (of the first four axles) of the Great Western 'Kings' was adopted, to the disadvantage of the overall design, no less than 8 ft. separating the first and second coupled axles. The cylinder disposition, the two inside over the leading bogie axle and two outside over the trailing axle, was the same as that on the G.W.R. four-cylinder engines, but they were given a slight slope (so avoiding the strange 'King' bogie design), and four sets of valve gear were provided. With cylinders 16¼ in. × 28 in., 6 ft. 7 in. coupled wheels and 250 lb. per sq. in. boiler pressure the tractive effort was 40,300 lb., the same as that of the 'Kings', all these dimensions being identical.

The underpart of the firebox resembled that of the Gresley 'Pacific' described in an earlier chapter) but the top was of Belpaire form. It furnished 190 sq. ft. heating and 45 sq. ft. grate area. This was united with a lengthy coned barrel increasing in diameter from 5 ft. 9 in. to 6 ft. 4¾ in. A moderate degree of superheat was provided by 16 flues (we use the term 'flue' in preference to 'element' in this context, as each of the former contained three of the latter).

If Stanier was indebted to Churchward, through Collett, and to Gresley for certain of the above-noted features, he was no less indebted to Hughes of the Lancashire & Yorkshire for the remarkable, and unique, form of the trailing end of these engines. The main frames were bifurcated at the rear of the coupled wheels, one member passing behind, and the other member outside, the trailing wheels, which were arranged on a Bissel truck.

The lengthy coupled wheelbase—15 ft. 3 in.—raised, in an acute form, the problem associated with the 'Pacific' wheel arrangement of bridging the space between the tube plates. The back tube plate was brought forward to some extent, even in the earlier engines, by the form of firebox already mentioned, but the front tube plate, though placed within the first barrel course, as in Great Western practice, was not recessed to any extent, with the result that the tubes were 20 ft. 9 in. long, and the total wheelbase and length over buffers—37 ft. 9 in. and 49 ft. $1\frac{1}{2}$ in.—compared unfavourably with Gresley's 'Pacific', in which the corresponding measurements were 35 ft. 9 in. and 43 ft. $11\frac{1}{4}$ in.

Stanier's first 'Pacific' turned the scale at $104\frac{1}{2}$ tons, of which 66 tons were available for adhesion.

In 1935 a new design of boiler was fitted to one of these engines (subsequently adopted for the class) with 32 flues, the small tubes being enlarged from $2\frac{1}{4}$ to $2\frac{3}{8}$ in. in diameter and reduced in number from 170 to 123. Matters were further improved by the provision of a combustion chamber, which reduced the length of the tubes to 19 ft. 3 in. and increased the firebox heating surface to 217 sq. ft.

Stanier's second series of 'Pacific' locomotives were of a more compact design. The coupled wheelbase was shortened by the abandonment of the 'King' dimension between the first and second coupled axle, the distance between these two being reduced by 9 in. to 7 ft. 3 in., which also separated the second axle from the third. The outside cylinders were moved forward and occupied a position just ahead of the trailing bogie wheels, instead of being over them, with the corollary that the inside and outside connecting rods were no longer of equal length. If Great Western practice was abandoned in these directions, it was adopted for the first time in another by the use of only two sets of valve gear for the four valves, though in this case the inside valves (as opposed to outside in Great Western practice) were actuated by rocking shafts.

The diameter of the piston valves was increased, compared with the earlier engines, by one inch to nine inches. The tractive effort was reduced to 40,000 lb. by an increase of $\frac{1}{4}$ in. in cylinder diameter and the adoption of 6 ft. 9 in. coupled wheels—a North Western dimension adopted by George Whale in his 'Precursors' of 1904, and perpetuated by Bowen Cooke and Fowler in the 'Claughtons' and 'Royal Scots'.

The boiler was considerably improved. The distance between the tube plates remained as in the re-designed boilers of the first series at 19 ft. 3 in. The barrel was now coned about its axis and the degree of superheat was raised by the provision of 40 flues, while the number of small tubes (still $2\frac{3}{8}$ in. in diameter) was increased to 129. The firebox was longer, the heating surface and grate being increased to 230 and 50 sq. ft. respectively.

With the shortening of the wheelbase and the provision of a combustion chamber, the smokebox was brought further forward, the front end now occupying a position virtually over the front bogie axle instead of being set considerably behind it as in the earlier engines. Compared with the original 'Pacifics' the overhang at the back (notwithstanding the longer firebox) was reduced by two inches. The front overhang was $4\frac{1}{2}$ in. more so that, with the reduction of the wheelbase, the total length, 48 ft. 7 in., was $6\frac{1}{2}$ in. less.

The earlier engines of this second series were provided with a streamline casing to work the 'Coronation Scot'. This was subsequently removed, but the engines retained the pared-down smokebox at the leading end, which rendered them easily distinguishable from their successors. The streamlined engines weighed 108 tons 2 cwt., of which 67 tons 2 cwt. was carried by the coupled wheels. The corresponding figures for the non-streamlined engines were 105 tons 5 cwt. and 66 tons 19 cwt. Subsequent engines of this series had double chimneys.

Ivatt

In the final version, built under the chieftainship of H. G. Ivatt, son of H. A. Ivatt of Great Northern fame, the trailing truck was redesigned. The double frames were abandoned in favour of two-inch bar frames riveted to the main frames behind the coupled axle and passing under the grate inside the trailing wheels after the American fashion, the trailing wheels themselves being attached to a steel casing of 'Delta' form with outside bearings. In Ivatt's engine the superheating surface was still further increased, by 93 to 949 sq. ft., in marked contrast to the pioneer engines which had only 370.

This final chapter of L.M.S. locomotive development may be fittingly concluded, as it began, with engines of the 2–6–0 wheel arrangement; though neither of the two classes to be noted (so far as the writer is aware) ever ran on the L.M.S. as such. They were, however, genuine L.M.S. products and were allocated to the London Midland Region of British Railways.

The first, produced under H. G. Ivatt, were general purpose engines intended for secondary duty. With cylinders 16 in. × 24 in., 5 ft. coupled wheels, a firebox 5 ft. 11 in. long, providing 101 sq. ft. heating surface and 17·5 sq. ft. grate area, and a total weight of 47 tons 2 cwt. (13 tons 11 cwt. maximum axle load), these engines were of quite moderate power. A noteworthy feature was that the barrel was coned about its axis as in the later 'Pacifics'. A companion 2–6–2 tank carrying the same boiler was a more compact version of Stanier's engines of this wheel arrangement, the length over buffers and total wheelbase being reduced from 41 ft. 11¾ in. to 38 ft. 9½ in. and from 33 ft. 3 in. to 30 ft. 3 in. respectively. The cylinders of the new engines were 16 in. by 24 in., compared with 17½ in. × 26 in. in their predecessors, the firebox was of the same length, and the boiler was no shorter but the water capacity was reduced from 1,500 to 1,350 gal., the coal capacity remaining at three tons.

The final 'Mogul' design on the L.M.S. presented a remarkable appearance with the very high running plate attached to the boiler, the 'V'-fronted cab, double chimney and external regulator rod. The boiler, which was of the pattern used on the 2–6–4 tanks, tapered, on its upper line, from 4 ft. 9⅛ in. to 5 ft. 3 in. The firebox provided 131 sq. ft. heating surface, the grate area was 23 sq. ft. and the boiler pressure 225 lb. per sq. in. The 17½ in. × 26 in. cylinders were provided with 10-in. valves—a very generous dimension—and the coupled wheels were 5 ft. 3 in. in diameter (Plate No. 44).

Comparison between these last two 'Moguls' with the first of Stanier's design affords a good illustration of the changes which had taken place during this last chapter of L.M.S. history and of the influence which the practice of this group was to have on British locomotive design in the final period.

III LONDON AND NORTH EASTERN

The locomotive history of the London & North Eastern Railway falls into two periods—that of nearly twenty years, during which Nigel Gresley was in charge, and that of under five, which witnessed the substantial changes in practice for which E. Thompson and his successor, A. H. Peppercorn, were responsible.

Gresley: Middle-range Duties

The most important feature of the Gresley period was the development of the large engine incorporating a wide firebox, a plan initiated by Ivatt in his 'Atlantics' of 1902 and perpetuated by Gresley himself in his pioneer 'Pacific' of 1922. Before going in to this line of development, however, a word should be said concerning Gresley's activities in providing engines of medium capacity. These consisted partly in the multiplication (with modifications) of engines derived from companies other than the parent Great Northern (of which Gresley had been the locomotive chief for over ten years before grouping)—including Great Central 'Directors' and 4–6–2 tanks, Great Eastern 4–6–0s and 0–6–2 tanks, and North Eastern 0–6–0s—and partly in the evolving of new designs. As regards the first, Gresley was far less ruthless than his opposite number on the L.M.S. and is known to have had a real appreciation of other than Doncaster products.

New designs of this power range comprised a 4–4–0 and a 4–6–0 with 5 ft. 6 in. diameter boilers—a dimension identical with that of Ivatt's large 'Atlantics' and Gresley's own 2–6–0s (of the middle series) and 2–8–0s. In the 4–4–0s advantage was taken of front axle drive to place all the cylinders in line below the smokebox. In the 4–6–0s the first coupled axle was used for the inside cylinder, and these engines provide the only example of divided drive being adopted by Gresley in the entire range of his three-cylinder designs. The length of the boiler was 11 ft. 5⅝ in. in the 4–4–0s and 13 ft. 6 in. in the 4–6–0s. The firebox of the former was 9 ft. long, providing 171·5 sq. ft. heating surface and 26 sq. ft. grate area. Corresponding dimensions for the 4–6–0s were 10 ft. 0½ in., 168 sq. ft. and 27·5 sq. ft. respectively. Accommodation for a firebox of this length was facilitated by the adoption of an unusually long wheelbase with 9 ft. between the second and third coupled axles.

Both the 4–4–0s and the 4–6–0s had 6 ft. 8 in. coupled wheels, while 17 in. × 26 in. cylinders in the former case, 17½ in. × 26 in. cylinders in the latter, and respective boiler pressures of 180 and 200 lb. per sq. in. gave tractive efforts of 21,556 lb. and 25,380 lb. The 4–4–0s (of the first series) weighed 66 tons and the 4–6–0s 76 tons 13 cwt., the respective adhesion weights being 42 tons and 53 tons 13 cwt. A similar boiler, cut to length and with a restricted firebox depth because of the presence of a coupled axle beneath it, was provided for the Great Eastern 4–6–0 '1500' class. It differed from those previously described in having a vertical throat plate which, with a firebox of practically similar length, gave a grate area of 31 sq. ft. but, because of the decreased depth, only 154 sq. ft. heating surface—14 sq. ft. less than on the Gresley engines; the barrel was about 1 ft. shorter.

A third class of engine in this power range was the 2–6–0 'K4', but these may be regarded as the adaptations of previous designs to the special conditions of the West Highland line, rather than the production of a new design, and may be somewhat briefly dismissed accordingly. A boiler of the pattern fitted to the second series of 2–6–0s, 5 ft. 6 in. in diameter, was combined with three-cylinder propulsion and 5 ft. 2 in. coupled wheels.

The 'Pacifics'

The development of the 'Pacifics,' culminating in the epic 'A4s' of 1935, is an oft-told story, which, notwithstanding its significance in British locomotive history, need not be repeated in detail here. It will be sufficient to recall that, as the result of the locomotive exchanges between Doncaster and Swindon in 1925, one of the 'Pacifics' was provided with long-travel valves and another also with boiler pressure raised by 40 lb. to 220 lb. per sq. in., the cylinder diameter being reduced correspondingly. These engines and others similarly altered were classified as 'A3'. The improvements both in performance and coal consumption were manifest and in the redoubtable 'A4s' Gresley went a stage further in raising the boiler pressure to 250 lb. per sq. in., reducing the cylinder diameter still further to 18½ in. Another improvement was the provision of a combustion chamber (in addition to that already furnished by design of the front end of the firebox already described in connection with the pioneer 'Pacific' of 1922) and the consequent shortening of the tubes by 1 ft. and an increase in the firebox heating surface from 215 to 231·2 sq. ft.

If Gresley was willing to learn from Great Western practice in the matter of steam distribution, he was by no means an uncritical disciple, and adhered to his conviction of the advantage of a high degree of superheat. Indeed, so far from adopting the low degree of superheat characteristic of Great Western practice, he stepped up the superheated volume of his 'Pacifics' subsequent to the Swindon-Doncaster trials by increasing the number of elements from 32 to 43. The 'A4s', which won immortal fame for the way in which they handled the 'Silver Jubilee' and subsequent 'Coronation' high-speed trains, are undoubtedly among the greatest locomotives ever constructed in Great Britain.

2–8–2s

The Gresley regime was also notable for the exploitation of the 2–8–2 type, and it may be regretted that the absence of suitable traffic-handling facilities in one case, and a change of policy due to cicumstances in no way reflecting on the design in the other, prevented a more extensive adoption of a type so admirably fitted for the haulage of heavy trains—goods and passenger alike—in this country.

The great advantage of this type of locomotive is, of course, that it enables the potentialities of a large boiler to be fully realized in practice by the provision of sufficient adhesive weight. In the course of our survey we have encountered designs in which boiler development has outrun the capabilities of 'the engine'. Several examples exist of the converse. In the 'Mikado' type each is happily balanced.

Gresley had in his 'Pacifics' a boiler admirably adapted for the purpose. In his first 2–8–2 engines, which appeared in 1925, this boiler was placed over wheels 5 ft. 2 in. in diameter. The four coupled axles were disposed over a wheelbase of 18 ft. 6 in., the distance between the first and second coupled axles being 6 in. greater than that between the other two to allow the inside connecting rod to clear the front axle. All the cylinders drove the second coupled axle. These engines weighed exactly 100 tons, of which 71½ tons rested on the coupled wheels. Only two were built, but it seems clear that they would have been a valuable acquisition to the locomotive stock of the L.N.E. had proper facilities for the handling of trains commensurate with their power been available.

Gresley's second design of this wheel arrangement appeared some nine years later. In this case 6 ft. 2 in. coupled wheels were used and the boiler was provided with a longer firebox, furnishing 50 sq. ft. grate area and 237 sq. ft. heating surface. The barrel was of the same dimensions as in the 'Pacifics', with a 43-element superheater. The first engine *Cock o' the North,* had poppet valves and rotary gear, but the second, *Earl Marischal,* was provided with Walschaerts gear, derived motion for the inside cylinder, and piston valves. Both were partially streamlined. Subsequent engines of the class, which were provided with longer combustion chambers increasing the firebox heating surface to 253 sq. ft., had the same wedge-shaped front as the 'A4s', but behind the smokebox the natural lines of the steam locomotives were allowed to appear.

The three cylinders, all of which drove the second coupled axle, were 21 in. × 26 in., which, with 220 lb. per sq. in. boiler pressure and coupled wheels of the size above-mentioned, gave a tractive effort of 43,462 lb. As might be expected, they proved themselves substantially more powerful than the 'Pacifics' on the Edinburgh–Aberdeen line for which they were designed and on which they worked. The pioneer engine weighed 110¼ tons, the coupled wheels, equally disposed over a wheelbase of 19 ft. 6 in., bearing 80 tons 12 cwt. This was the largest engine of the conventional type produced by Gresley, and it is much to be regretted that they were not extensively multiplied.

2–6–2s

Another wheel arrangement insufficiently appreciated in this country is the 2–6–2, and here again Gresley provides two notable examples (one markedly successful). The first of these followed by some two years the pioneer 2–8–2 passenger engine just described and, here again, the coupled wheels were 6 ft. 2 in. in diameter. The adaptation of the 'Pacific' boiler was a simple matter and, indeed, the substitution of a pony truck for a bogie conveniently reduced the space to be bridged between the tube plates; this was now 17 ft. The firebox was of the same size as that fitted to the 'Pacifics.'

The coupled wheelbase was 15 ft. 6 in., 7 ft. 3 in. separating the first and second axles (as on the 'Pacifics') and 8 ft. 3 in. the second and third. Advantage was taken of the smaller wheel to place the rear coupled axle a few inches nearer the firebox and the wheelbase between the rear coupled and trailing wheels was shortened by 3 in. to 9 ft. 3 in. At the other end the substitution of a pony truck for a bogie enabled a further 2 ft. 10 in. to be saved, with the result that the total wheelbase of the new engines was 33 ft. 8 in.—3 ft. 1 in. shorter than that of their express counterparts. A 43-element superheater was provided. The cylinders were 18½ in. × 26 in.; the boiler was pressed to 220 lb. per sq. in., giving a tractive effort of 33,730 lb., and 65 tons 12 cwt. of the total weight of 93 tons 2 cwt. were available for adhesion. In all, the design exhibited a remarkable combination of power and compactness. These engines performed admirably and, 6 ft. 2 in. coupled wheels notwithstanding, were capable of very high speeds.

Gresley's second 2–6–2 design, of which two engines appeared in 1941, was a light version of the locomotive just described. The barrel, the second course of which was coned about its axis, tapered from 4 ft. 8 in. to 5 ft. 4 in. The firebox, which was of the same formation as that used in Gresley's larger engines, provided 152 sq. ft. heating surface (increased in the second engine by a Nicholson syphon) and 28 sq. ft. grate area. The 5 ft. 8 in. coupled wheels were disposed over a wheelbase of 12 ft. 10 in. and carried 48 tons 11 cwt. of a total weight of 70 tons 8 cwt. The maximum axle load of 17 tons greatly increased the availability of these engines. With the three 15 in. × 26 in. cylinders and 250 lb. per sq. in. boiler pressure, the tractive effort was 27,420 lb. The type was not perpetuated, but whether this was due entirely to a change of locomotive chiefs (the engines appeared towards the end of Gresley's career and were indeed his last design), it is hard to say.

Water Tube Boiler

Two further Gresley designs, each represented by a single locomotive, now call for notice. The first combined a water-tube boiler with a 'Pacific' chassis, substantially modified to carry it. A single trailing axle would have been insufficient to bear the additional weight, so a further axle was provided. Lateral movement was given to the first by Cartazzi axle boxes (as on the

41. Pannier Tank. Typical example of Great Western individuality, 1928
42. Stanier's three-cylinder tank, L.M.S., 1934

43. Thompson's general utility 4–6–0 for the L.N.E.R., 1942
44. The last L.M.S. 'Mogul' by H. G. Ivatt, 1947

'Pacifics') and to the second by a Bissell truck—an extraordinary arrangement which led the purists (or pedants?) to describe the locomotive as of the 4-6-2-2 type.

The other modification of the 'Pacific' chassis consisted of the substitution of two high-pressure and two low-pressure cylinders, with divided drive, for the three-cylinder single-axle propulsion. This was undoubtedly because of the high working pressure of 450 lb. per sq. in., rendered possible by the design of the boiler. The high-pressure cylinders, originally 12 in. and later 10 in. × 26 in., were inside and drove the leading coupled axle; the low-pressure cylinders, 20 in. × 26 in., of Gresley's standard design, actuated the middle axle. Only two sets of valve gear were fitted but a half-link introduced into the connecting mechanism enabled the cut-off of high- and low-pressure cylinders to be varied at the driver's discretion.

This engine, which turned the scale at 103 tons 12 cwt., was subsequently rebuilt with three high-pressure cylinders 20 in. × 26 in. and a standard boiler as fitted to the 2-8-2 express engines.

Garratt

The last Gresley engine calling for mention is the remarkable Garratt 2-8-0+0-8-2, which appeared in 1925. Here an enormous boiler, of greater diameter than any other in this country before or since, was slung between two chassis of the designer's 2-8-0 three-cylinder mineral engines.

Particular interest attaches to the boiler, which was of the simplest possible formation with parallel barrel and round-topped firebox having vertical sides. In its simplicity and its proportions this boiler might well be described as a designer's dream. The maximum diameter of the barrel was 7 ft., and the length of the barrel was 13 ft. There were 275 fire-tubes and 45 flues, which gave 2,757 sq. ft. heating surface. The outer firebox measured 9 ft. 4 in. and provided 237 sq. ft. more, which, with 646 sq. ft. superheating surface, brought the total to the impressive figure of 3,640 sq. ft.—all of it, because of the moderate length of the tubes, of real value. The grate area was 56·4 sq. ft. With six 18½ in. × 26 in. cylinders, 4 ft. 8 in. coupled wheels and 180 lb. per sq. in. boiler pressure, the tractive effort was 72,940 lb.

After Gresley

The concluding phase of London & North Eastern locomotive history is marked by important changes introduced by Gresley's successors. One of these was prompted by a desire to eliminate a source of trouble in his 'Pacifics.' The derived motion tended to overrun at high speed and threw on the inside big-end a strain which it was not always able to withstand. Another arose from the conviction that, for all but the largest engines, the elimination of the third cylinder would be an advantage.

The result of the latter is exemplified in the production of a serviceable and highly competent 4-6-0 by E. Thompson in 1942 (Plate No. 43). The boiler, 5 ft. 6 in. in diameter, and all but 14 ft. between tube plates, with a large firebox providing 168 sq. ft. heating surface and 27·5 sq. ft. grate area, resembled that used on Gresley's three-cylinder 4-6-0s, but the layout of the machinery was much simplified by the use of two outside cylinders only (with motion, of course, similarly situated). These engines weighed 71 tons 3 cwt. with 52½ tons over the coupled wheels. Tractive effort, derived from 20 in. × 26 in. cylinders, 6 ft. 2 in. coupled wheels and 225 lb. per sq. in. boiler pressure, was 26,878 lb. The maximum axle load was 17 tons 15 cwt. and these engines provided a far simpler solution than Gresley's lightweight 2-6-2 of the problem of meeting mid-range demands of all descriptions, and were an undoubted acquisition to the L.N.E. locomotive stud.

The writer finds it less easy to view with favour certain other aspects of Thompson's work—particularly the reconstruction of Gresley's admirable 2–8–2s as (it must be said) somewhat contorted 'Pacifics'. The plan adopted for the transformation was, substantially, to leave the last three coupled axles in position and connect the outside driving mechanism, which had originally operated the second coupled axle, to what was originally the third and was now the middle axle. The place formerly occupied by the leading coupled wheels was now taken by the outside cylinders and the bogie was situated ahead of them. The inside cylinder was placed in the conventional position over the bogie and drove the leading coupled axle, separate motion being provided for the valve. The coupled wheelbase, of 13 ft., was somewhat short to take the 66 tons placed over the coupled wheels, and that between the first coupled and second bogie axles (8 ft. 2 in., compared with 5 ft. 6 in. on the Gresley 'Pacifics') was inordinately long. The cylinder diameter was reduced by one inch to 20 in. and the boiler pressure raised by 5 lb. per sq. in. to 225, the tractive effort being now 40,318 lb.

This conversion took place in 1943. In the following year some 'V2s' under construction were produced as 'Pacifics' conforming with this plan, and in 1945 Gresley's original *Great Northern*—of all engines—appeared so modified, though the original coupled wheelbase (incapable of being reduced to the limits prevailing in the other engines by reason of the 6 ft. 8 in. wheels) remained unaltered.

A. H. Peppercorn, Thompson's successor, also adopted divided drive in the two designs of 'Pacific' engines for which he was responsible. The first of these, with 6 ft. 2 in. wheels, appeared in 1947, and the second, with 6 ft. 8 in. wheels, in 1948. An inside-cylinder connected to the first coupled axle entailed, or at least rendered convenient, a longer wheelbase than the Gresley plan under which all the cylinders occupied the conventional position, but the extension was not so great as in the earlier Thompson engines, the distance between the rear bogie and leading driver being 5 ft. 9 in.—only 3 in. more than on the Gresley engines. Thompson's very long smokebox was, however, perpetuated, with the result that the front end was level with the leading bogie axle, instead of being well behind it as in Gresley's design.

The boilers of both the Peppercorn series had the shorter barrel measuring 17 ft. between tube-plates and the longer firebox furnishing 50 sq. ft. grate area. The boiler pressure was 250 lb. per sq. in. The new form of divided drive and the provision of a separate valve-gear for each cylinder were good features, but the Peppercorn engines never superseded the 'A4s' and it is significant that, when any exceptional task had to be performed, one of the latter was usually chosen for the purpose.

As has been indicated, this review of London & North Eastern locomotive practice makes no claim to be complete. Mention has not been made, for example, of Gresley's experiments with boosters, nor of his 2–6–2 tanks or Thompson's 2–6–4 tanks, nor of the latter's rebuilding of one of Robinson's 0–8–0s with tanks and bunker for shunting purposes, but none of these contribute materially to the general picture.

The great feature of L.N.E. practice was the exploitation of the 'big engine'. In this it took the leading position and it was due to Gresley's foresight that, when this chapter of locomotive practice was concluded, the group over which he had for so long presided surrendered to British Railways a far greater number of large engines than any of the others—indeed than all the others put together.

IV THE SOUTHERN

Measured in terms of volume, the demands for locomotive power on the Southern was far less than on either of the much larger northern groups or—by reason of the extensive adoption of electric traction, considerably augmented during the period under review—on the Great Western, which, in size, it more closely resembled. Nevertheless, designs of considerable interest and merit were produced, and it is to the Southern in its latter days that we must look for the most original engines, produced on a commercial scale, ever to run on British railways since the Stephenson engine reached its present form.

Maunsell-Eastleigh

Southern locomotive practice falls into two well-marked epochs—presided over by R. E. L. Maunsell and O. V. S. Bulleid respectively. The former period witnessed an unresolved conflict between Ashford and Eastleigh practice, with Brighton not entirely submerged. The first-named was represented by the perpetuation of the South Eastern 'Moguls' and corresponding tank locomotives for middle-range duties; Eastligh provided the groundwork for express locomotives; while an original 0–8–0 shunting tank owed its boiler, if nothing else, to Brighton practice.

Maunsell's first express locomotives—the 'King Arthurs'—were direct derivatives of Urie's South Western creations, the features of which have been noted in an earlier chapter. But the boiler pressure was raised to 200 lb. per sq. in., the cylinders were reduced to 20 in. × 26 in., long travel valves were used and the draught arrangements were overhauled. In the result a much more lively engine was produced, capable of dealing successfully with the Southern trains of the period. The Urie 5 ft. 7 in. 4–6–0 goods engine was perpetuated, with Maunsell features, in a series built in 1927.

The demand for greater power for express duties was met by the production of the 'Lord Nelson' class in 1926. Here the boiler diameter was increased to 5 ft. 9 in., and the length between the tube plates was 14 ft. 2 in. The coning of the first boiler course—a 'King Arthur' feature derived from Urie—was eliminated. The Belpaire firebox, 10 ft. 6 in. long, furnishing 194 sq. ft. heating surface and a grate area of 33 sq. ft., had slightly curved plates and the grate was partly level and partly sloped—both these features being reminiscent of Swindon practice. The four cylinders, 16½ in. × 26 in., were arranged virtually, but not completely, in line, after the manner adopted by Hughes on the Lancashire & Yorkshire and the L.M.S., and the drive was divided, those inside actuating the leading and those outside the middle coupled axle.

The most unusual feature of these engines was the crank setting, the inside and outside cranks of the engine being set at 135 deg. instead of the customary 180 deg. The arrangement had been tried on one of the earlier Drummond 4–6–0s, with, it was said, very satisfactory results, and was first employed on a North Staffordshire Railway 0–6–0 tank in 1922. It involved the loss of the perfect balance secured by the more usual arrangement but, with the eight power impulses for each revolution of the driving wheels, provided a more even turning movement and a more even pull on the fire. In other words, it carried a stage further the advantages secured by the three-cylinder engine with 120-deg. crank setting over both its two-cylinder and four-cylinder rivals. As in so many problems associated with locomotive design, this was a matter of give and take—of striking a balance between advantages and disadvantages—and here, as elsewhere, the history of the steam locomotive does not admit of an unequivocal verdict.

The resetting of the cranks of one of the engines in accordance with the more usual plan, with indeterminate results, was one of the expedients adopted to improve these not wholly satisfactory engines in their original form. Another was the provision of a very large centrally coned boiler with a round-topped firebox coned from front to back and furnished with a combustion chamber protruding into the barrel. This may have been carrying a good thing too far, because, with a barrel of large diameter, the firebox, necessarily narrow at the grate, substantially widens at the centre line, and this enforced difference performs the same office as the combustion chamber in a locomotive provided with a wide grate, where the position as between the respective widths is, of course, reversed.

An Outstanding 4–4–0

The most noteworthy design of express locomotive produced by Maunsell—and the most successful in terms of work extracted from a given weight—was unquestionably that of the three-cylinder 4–4–0 'Schools'. Here a firebox of 'King Arthur' proportions was united with a shorter boiler of similar diameter. The three cylinders, 16½ in. × 26 in., were arranged in a line below the smokebox and drove the leading coupled axle. With these dimensions, 6 ft. 7 in. coupled wheels and 220 lb. per sq. in. boiler pressure, the tractive effort was 25,130 lb.—a high figure for a 4–4–0, of which the 'Schools' are usually regarded as the supreme examples. Dimensionally they were approached by Wilson Worsdell's 'R1s' of 1910, which had a fractional advantage in boiler diameter and pressure, the same adhesion weight (42 tons), but only 27 as against 28·3 sq. ft. grate area. Gresley's 4–4–0s were also of comparable capacity. But few would quarrel with the awarding of the palm to the 'Schools', which may fairly be regarded as Maunsell's greatest achievement, none the less remarkable in that the engines were planned for the restricted loading gauge of the Hastings line—to which their cabs, inward sloping from the waist line, bore visible witness. Maunsell also achieved great success in the rebuilding of the Wainwright 4–4–0s (Plate No. 17) and in new engines conforming to this pattern (Plate No. 39).

Ashford

The Ashford tradition was exemplified by the 2–6–0s and 2–6–4 tanks constructed for middle-range duties. The characteristics of these engines were fully analysed in a previous chapter, and need not be further described here; but it is no criticism of the admirable 'King Arthurs' to express regret that the same ideas were not embodied in a larger engine suitable for the haulage of the heaviest and fastest trains. A Maunsell boiler, tapering from 5 ft. 3 in. to 5 ft. 9 in. and 13 ft. between tube plates, with a 10 ft. Belpaire firebox giving some 32 sq. ft. grate area and narrowing from throat to back by 1 ft., would have constituted an admirable steam raiser for a 4–6–0 provided with, say, 20 in. × 28 in. cylinders and 6 ft. 6 in. coupled wheels. And, if traffic demands increased beyond the capacities of such an engine, they could still have been met by the application of the same features to a still larger machine—perhaps of the 4–8–0 type, a wheel arrangement known to have been contemplated by Maunsell himself, though for an engine of the 'Lord Nelson' variety adapted for goods traffic.

The Ashford engines appeared in various forms—tender and tank, two-cylinder and three-cylinder, and with 5 ft. 6 in. and 6 ft. coupled wheels. Seven of the eight possible combinations of these varieties were realized in practice—the exception being a two-cylinder tank engine with the smaller wheels—though all the 2–6–4 tanks with 6-ft. wheels had been converted to tender engines by the time the last constructed variety, a 2–6–4 tank with 5 ft. 6 in. wheels, appeared. In all, some 170 engines of these classes were built, forming a valuable nucleus of Southern motive power.

Another design which bore unmistakable Ashford characteristics was that of a neat inside-cylinder 0–6–0, though the absence of a coned boiler and outside motion rendered it very different from the locomotive just described. This example of the employment of an earlier classic British design had its counterpart in the 4–4–0 express locomotive built for the Southern in 1926, following the successful rebuilding of the earlier Wainwright engines of the same wheel arrangement.

Brighton

Brighton has already been credited with the origin of the 0–8–0 tanks (which were, in fact, constructed for shunting duties) but perhaps rather fancifully, for nothing characteristic of Brighton work was to be found therein save the boiler, and even this departed from the standard pattern, in having the smokebox directly riveted to the first barrel course without the interposition of a distance ring. This boiler was of moderate capacity—5 ft. in maximum diameter, 10 ft. 7 in. between tube plates, with a grate area of 18·4 sq. ft. Three cylinders, 16 in. × 28 in., of the type fitted to the 'Moguls', and each with its own motion, were used. The inside cylinder drove the second, and the outside cylinders the third, coupled axle. The distance between the first and second axles was greater than that between the second and third to allow good room for the inside machinery, after the manner of McIntosh's 0–8–0 mineral engines for the Caledonian Railway, but that designer's wide spacing between the third and fourth axles (to allow the firebox to be dropped between them) was not adopted and, indeed, would hardly have been practicable on an engine required to carry its own water and coal supplies and upon which, in consequence, the firebox had necessarily to terminate further forward. In fact, on the Southern engines the firebox occupied a position almost midway over the third coupled axle, an arrangement which permitted the provision of reasonably capacious tanks and bunker for a shunting engine, holding 1,500 gal. and 3 tons respectively.

With the cylinder dimensions stated, 4 ft. 8 in. coupled wheels, and 180 lb. per sq. in. boiler pressure, the tractive effort was 29,380 lb., while three-cylinder propulsion, coupled with the fact that the whole weight of the engine—71 tons 12 cwt.—was available for adhesion, rendered the best possible use to be made of the power capable of being exerted.

Bulleid-'Pacifics'

We now enter on what is unquestionably one of the most remarkable epochs of locomotive history in the country. This was due to the originality of O. V. S. Bulleid, who was in charge during this second (and last) period of the existence of the Southern Railway.

Bulleid came to the Southern from Doncaster and his initial design, that of the 'Merchant Navy Pacifics', derived its basic plan from Gresley, in that three high-pressure cylinders drove the middle axle and that a wide firebox was adopted. That said, all resemblance virtually ceased. The firebox, constructed of steel, was of Belpaire formation, somewhat steeply inclined on top. Below the centre line, L.N.E. practice was discernible, in that the wide portion was extended to the centre by a forward slope, the portion below being semicircular, with the rear coupled wheels beneath it. The tube plate was extended into the barrel to the maximum length compatible with the insertion of the inner firebox into the casing at the foundation ring, this form of construction being facilitated by the very pronounced outward slope given to the back plate from near the top to the bottom. The grate area was 48·5 sq. ft. and total heating surface 2,451 sq. ft., the superheater adding a further 822.

The highly-efficient firebox, which had two Nicholson syphons built into it, was united with a no less remarkable barrel, the second ring of which, for the first and only time in British locomotive history, was tapered down towards the firebox, the upper line of the barrel plate remaining horizontal. This was a particularly valuable feature where, as here, the firebox end had to accommodate a combustion chamber with its surrounding water space. Gresley, it will be recalled, adopted a central cone in similar circumstances, so did Stanier in his later 'Pacifics', but it was left to Bulleid to carry on the process to its logical conclusion, in complete contrast to the Churchward practice, where the motive was to provide extra steam space at the part of the boiler where steam was most rapidly generated.

The diameter of the 'Merchant Navy' boiler increased from 5 ft. 9¼ in. in front to 6 ft. 3½ in. at the rear, and the length between tube plates was 17 ft. The upward coning raised the boiler centre line at the front end and would have left very little room for a chimney had not the upper part of the smokebox (which housed a multiple orifice blastpipe) been given a pronounced downward slope—and very curious these engines would have looked had they not been shrouded in an overall casing. Another feature hidden from view was the setting-back of the smokebox in relation to the bogie. Indeed, the chimney occupied a position slightly aft of the second bogie axle. The writer has often regretted that the true lineaments of at least one of these engines was not allowed to appear. They would have shocked the conventional and given satisfaction to those whose interest in locomotive design is more than skin deep.

The position of the cylinders resembled that on the Gresley engines but the inside cylinder was brought rather nearer the driving axle. The valves and gear were wholly different. The former were arranged for outside admission, the exhaust being in consequence very direct. Perhaps the most unusual—and questionable—feature of the design was the method of actuating the valves. The gear for each was taken from a shaft to which motion was imparted by chain drive, the whole, including the inside cross-head and connecting rod, being enclosed in an oil bath, somewhat voracious in its demand for lubricant and apt to leak, with disastrous results on the slipping propensities of these engines.

The 6 ft. 2 in. coupled wheels were equally disposed over an ample wheelbase of 15 ft., which provided sufficient clearance for the clasp brakes with two shoes for each wheel—another unusual, if not quite original feature. The leading end was carried on a bogie with the short wheelbase of 6 ft. 3 in. and, at the other end, the firebox was supported by a truck of delta form with its axle 10 ft. distant at rail-level from the third coupled axle, the main frames passing under the firebox. The back end of the engine thus conformed to the usual American pattern (though without bar frames), in contrast to the expedients adopted by Stanier and Gresley.

A short piston stroke of 24 in., together with large-diameter valves (11 in.) and well designed exhaust passages, counteracted any tendency to sluggishness that might have been expected to arise from the restricted diameter of the coupled wheels, and these engines were among the swiftest ever to run in this country.

The tractive effort with 18-in. cylinders, 280 lb. per sq. in. boiler pressure—never before approached and only once equalled in British practice for a normal boiler—and the other dimensions already given amounted to 37,500 lb. The estimated weight was 92½ tons, with 21 tons on each of the coupled axles. In fact the engines turned the scale at 94¾ tons.

A lighter version of this design appeared a few years later in the 'West Country' class—subsequently augmented by the 'Battle of Britain' series—which had a maximum load of 18¾ tons, which was carried by each of the coupled axles. Piston stroke, coupled-wheel diameter, length between tube-plates, and maximum (though not minimum) boiler diameter remained the same, but the diameter of the cylinders was reduced to 16⅝ in. and of the valves to 10 in., heating and superheating surface to 2,122 and 545 sq. ft. respectively, and the grate area by over 10 sq. ft. to 38·23, and the total weight to 86 tons. These smaller dimensions enabled the engines to be used on most of the Southern routes where fast running was demanded.

Unlike Gresley's scaled-down 2–6–2s, these light 'Pacifics' (whose leading dimensions could have been provided on a far less elaborate design with a narrow firebox) were built in large numbers and set up a standard of performance little inferior to that of their larger counterparts.

Notwithstanding the remarkable achievements of both these classes, it would be less than frank to disguise the fact that some of their original features—notably the chain-driven valve gear and the oil bath—gave considerable trouble in operation. But it is a notable tribute to their designer that when, subsequent to the demise of the Southern group, they were rebuilt with conventional valve gear, the original boiler remained unaltered, though the pressure was lowered to 250 lb. per sq. in.

0–6–0

The vagaries of genius are unpredictable, and it is of interest to note that in his second design Bulleid adopted a long-established basic plan—one indeed characteristic of British practice many years before the changes demanded by traffic development at the end of the Victorian era. In other words, he adopted the 0–6–0 wheel arrangement with inside cylinders and valve gear—a plan dating back to John Ramsbottom's 'DX' class of 1858, and indeed beyond it.

With these 'Q1' engines Bulleid used the firebox pressings of his predecessor's 'Nelsons', which gave an unprecedented boiler diameter for an engine of its type of 5 ft. 9 in. and also a capacious firebox providing 170 sq. ft. heating surface and 27 sq. ft. grate area. The boiler, which was coned about its axis, was 5 ft. in diameter at the smokebox and the barrel was 9 ft. 9¼ in. long. With 19 in. × 26 in. cylinders, 5 ft. 1 in. wheels and 230 lb. per sq. in. boiler pressure, the tractive effort was 30,000 lb. The total weight was 51¼ tons and the maximum axle load (as in the smaller 'Pacifics') 18 tons 15 cwt.

If the general layout of these locomotives was conventional, their appearance certainly was not. No running plate was provided and the cab was perched high up on the end of the boiler. The boiler clothing did not follow the lines either of barrel or firebox but, in section, it took the form of an inverted horse-shoe. It was constructed in two stages, the second being slightly wider and higher than the first, the underside being flat. The smokebox also followed these lines and the chimney, placed over a multiple-jet blast-pipe, was an uncompromising stove pipe of considerable width. The general ensemble was not rendered less strange by the 'boxpok' form of steel wheel centre. In service these engines proved to be very powerful and they may surely be regarded as exhibiting the highest stage of development to which the classic British 0–6–0 was ever brought.

Unfinished Experiment

This completes the story of the Bulleid engines brought into service, and perhaps little need be said of the even more remarkable 'Leader' class, which was still in the experimental stage when the Southern became merged into British Railways. These amazing machines had two six-wheeled power bogies, each driven by three cylinders. The boiler was placed (slightly aslant!) between them and there was a cab at both ends. The changes wrought by nationalization put an end to the trials of the examples which had been built and the parts ready for the assembly of further examples were scrapped. In the circumstances their influence on locomotive practice in Great Britain was negligible. Nevertheless they are of interest and of importance as constituting an unfinished chapter in the history of the most original designer ever to be associated with the steam locomotive since the days when it assumed the basic form in the middle years of the nineteenth century.

Other groups were contemplating the production of locomotives of greater power than any previously built—though on conventional lines—and it is fitting that this chapter of British locomotive history, which began with the frustration of many promising lines of development on the part of the individual companies, should itself be brought to an end on a note of projects unrealized.

CHAPTER VIII

The End

HAVING regard to the events that were to follow, the production by British Railways of a dozen new standard classes of locomotive, all between 1951 and 1954, is a surprising phenomenon. Had it been appreciated that the steam locomotive was so near the end of its career, it can hardly be doubted that a few of the most successful of the pre-nationalization classes would have been selected for multiplication as a stop-gap and the observer of British locomotive practice would have been deprived of the opportunity of witnessing the last thoughts of locomotive engineers expressed in a series of classes bearing a strong family likeness to each other but widely diversified, so as to cover practically the whole range of operative duties.

An examination of the various classes should be prefaced with a few general observations.

The designers turned their backs on the high-efficiency engine, in the production of which the French excelled. Neither compounding nor feed-water heating was resorted to. Simplicity was the keynote. The principal aim was to produce a machine that would give reliable service over a long period with the minimum of attention. Rocking grates were incorporated and special attention was paid to accessibility. An outstanding advantage of the steam locomotive over other forms of power is its ability to continue to give useful service when not in the best of condition, and in these robust classes, fully representative of British engineering skill at its highest, this characteristic was exploited to the utmost.

All the standard classes were six-coupled save one—the 2–10–0 general utility engines, the first *class*, apart from that constructed for war purposes, of British locomotive to embody five-coupled axles—Holden's 'Decapod' and Fowler's banker being, of course, but solitary examples. The absence of any four-coupled design from the new range is hardly surprising, but the abandonment of the eight-coupled engine, which had been extensively used for mineral work over a period of sixty years, is less easily explained. Perhaps the stock of existing eight-coupled locomotives, which had been substantially augmented in the years preceding nationalization, was ample. Moreover, it is understood that the preliminary plans for the heavy mixed traffic locomotive originally incorporated the 2–8–2 wheel formation.

The other standard classes comprised three 'Pacifics' (one never multiplied), two 4–6–0s, three 2–6–0s, and three tank engines, one of the 2–6–4 and two of the 2–6–2 order.

Wide fireboxes were employed on the 'Pacifics' and the 2–10–0. The boiler characteristics of the remaining classes with narrow fireboxes were derived from Swindon practice, as modified by the L.M.S. The adoption of so elaborate a model, with its coned barrel and radial firebox plates (narrowing from the front to the back in the larger engines) contrasts with the austerity of the rest of the design; but it was probably justified by the more even circulation of the water thereby promoted and consequent longer life.

In every case save one—that of the largest 'Pacific'—only two cylinders were employed, placed, of course, outside with their attendant valve and gear, and, in all but the smallest engines, sharply inclined to clear the loading gauge operating over the British system generally. Hitherto, British practice had favoured the use of more than two cylinders on engines destined for the hardest duties.

Cabs were perched high in the manner reminiscent of early Churchward practice, and there was no attempt to disguise the height of the footplate by a downward prolongation of the sides to footstep level. The footplates were extended backwards over the tender which were also provided with protection for the engineman and with a set of steps giving access to the overhanging portion of the cab. Boilers with narrow fireboxes were no longer supported by the time-honoured expansion brackets at the sides but by an attachment fitted to the fabricated drag-box at the rear. Regulator rods were external.

Produced under the aegis of R. A. Riddles (member for mechanical and electrical engineering), R. C. Bond (chief officer, locomotive works) and E. S. Cox (executive officer—design), the new classes were more reminiscent of L.M.S. practice than that of any of the other groups. A detail worthy of mention in this connection is the resuscitation of the Horwich chimney, a distinguishing feature of Lancashire & Yorkshire many years earlier.

'Pacifics'

The first of the new classes to appear was the large two-cylinder 'Pacific'—class '7'—which made its debut in 1951. This was a less powerful machine than the existing locomotives of the same wheel formation on the northern lines; but was the most powerful unit hitherto produced in Britain with only two cylinders, and may be regarded as the highest combination of capacity and simplicity yet realized in this country (Plate No. 51).

20 in. × 28 in. cylinders, somewhat sharply inclined, drove 6 ft. 2 in. coupled wheels. The trailing wheels were accommodated on a 'delta' truck—a more modern arrangement than those originally adopted in British engines of this type. The first barrel course was parallel; the second was coned about its axis, the maximum and minimum diameters being 5 ft. 9 in. and 6 ft. $5\frac{1}{2}$ in. It contained 136 $2\frac{1}{8}$-in. and 40 $5\frac{1}{2}$-in. tubes, 17 ft. long. The firebox, which was of Belpaire formation at the top and followed Gresley's and Stanier's practice in having the wide portion extended at an angle from the grate to the boiler centre line with a hemispherical portion accommodating the combustion chamber beneath it, was 7 ft. long and 7 ft. 9 in. and 7 ft. 4 in. wide at grate level at front and rear respectively. It furnished a grate area of 42 sq. ft. and 210 sq. ft. heating surface. The engine turned the scale at 94 tons, of which $60\frac{3}{4}$ tons, equally distributed, rested on the coupled wheels.

The second 'Pacific'—class '6'—was a smaller version of the locomotive just described. The cylinders were $\frac{1}{2}$-in. less in diameter. The boiler, of a maximum diameter of 6 ft. 1 in. contained 5 fewer large and 18 fewer small tubes. The firebox, which was 3 in. shorter and 9 in. narrower in front, furnished 195 sq. ft. heating surface and 36 sq. ft. grate area. Boiler pressure was lower by 25 lb. per sq. in. These changes brought the total weight down to $88\frac{1}{2}$ tons, with an adhesion weight of $56\frac{1}{4}$ tons and a maximum axle load of under 19 tons—figures which gave these engines a wider range of operation than their larger confrères.

The third 'Pacific' design—that of class '8'—represents the efforts of British Railways to provide an express locomotive of maximum capacity, consonant with the limits imposed by three

45 & 46. Great Western four-cylinder 4–6–0s in action, of *Castle* and *King* classes, introduced 1923 and 1927

47. Intermediate class of B.R. 2–6–0s
48. Tank version of the same

49. The largest B.R. 2–6–0
50. B.R. version of Stanier's 4–6–0 L.M.S. mixed traffic locomotive

51. Standard B.R. express locomotive
52. The ultimate development, incorporating wide firebox over coupled wheels

coupled axles and the loading gauge and weight restrictions. The power required involved departure from the 'all outside' position of cylinders, valves and gear; but the inside machinery was kept down to the minimum by the adoption of rotary cam gear and poppet valves. The three cylinders measured 18 in. × 26 in., the leading coupled axle taking the drive of the inside cylinder, after the manner of later L.N.E. practice. As on the other B.R. 'Pacifics,' the coupled wheels were 6 ft. 2 in. in diameter. The barrel and tube arrangements were similar to those on the larger two-cylinder engines; but the firebox was 9 in. longer, giving an enhanced heating surface of 226 sq. ft. and a grate area of 48·6 sq. ft. The adhesion weight went up to 66 tons (equally distributed over the three axles) and the total weight to 101¼ tons. Unfortunately, this engine—a solitary example of its class—appeared at a time when the steam locomotive as the source of power on the heaviest duties was doomed, and there is little evidence of its potentialities.

4–6–0s

The two 4–6–0s—of classes '5' and '4'—were of very similar design, though of different capacity. Class '5' (Plate No. 50) was a B.R. version of Stanier's famous engines, the boilers being of identical dimensions. The coupled wheels, 6 ft. 2 in., were slightly larger. The smaller engines had 5 ft. 8 in. coupled wheels. Both had a piston stroke of 28 in., the respective diameters being 19 in. and 18 in. The maximum boiler diameter of class '5' was 5 ft. 8½ in., that of class '4' was 5 ft. 3 in., the second courses being tapered upwards. 28 5⅛-in. tubes were provided on the larger engines and 21 on the smaller, the respective number and diameters of the small tubes being 151 of 1⅞-in. and 157 of 1¾-in. The fireboxes of class '5' were 9 ft. 2 13/16 in. long and furnished 171 sq. ft. heating surface and 28·65 sq. ft. grate area. Corresponding dimensions of the smaller engines were 8 ft. 6 in., 143 and 26·7 sq. ft. The boiler pressure in each case was 225 lb. per sq. in. The total weights were 76 tons and 67 tons 18 cwt. Class '5' carried just over 58 tons on its coupled wheels, class '4' 6½ tons less. Maximum axle loads were 19 tons 14 cwt. and 17 tons 5 cwt.

2–6–0s

The three classes of 2–6–0 locomotive were distinguished by a similar gradation of power. Class '4' (Plate No. 49) was based on the L.M.S. class '4' and in lateral boiler dimensions resembled the smaller 4–6–0 already described. Class '3' (Plate No. 47) was modelled on the Great Western 2–6–2 tank, but had a shorter boiler, while class '2' was intended to provide a thoroughly modern locomotive for light branch duties. The first two classes had 17½ in. × 26 in. cylinders and 5 ft. 3 in. coupled wheels. In class '2' the corresponding dimensions were 16½ in., 24 in. and 5 ft. In each case the length between the tube plates was the same—10 ft. 10½ in. The number of 5⅛-in. tubes provided was respectively 24, 18 and 12, the corresponding number of small tubes being 154, 145 and 162—all of 1⅝-in. diameter. Class '4' had a firebox 7 ft. 6 in. long. That of class '3' was 6 in. and that of class '2' 1 ft. 7 in. shorter. Heating surfaces were 131, 118 and 101 sq. ft. and grate areas 23, 20·35 and 17·5 sq. ft. respectively. Class '4' carried a working pressure of 225 lb. per sq. in., the others 200. Class '4' weighed 59 tons 2 cwt., class '3' 57½ tons and class '2' 49¼, the respective maximum axle loads being 16¾, 16¼ and 13¾ tons.

Tank engines

The three tank engines can be somewhat briefly dismissed.

The most powerful—the 2–6–4—was virtually an adaptation of the Class '4' 4–6–0 for short range duties. Laterally the boiler dimensions of both classes were the same, though the substitution on the tank engine of a pony truck for a bogie at the leading end involved a shortening of the boiler, the number and diameter of the tubes being retained. Cylinder dimensions and coupled wheel diameter were identical; so was the boiler pressure—225 lb. per sq. in. 53 tons of the total weight of 86 tons 13 cwt. were available for adhesion, the comparatively high maximum axle load being 17 tons 19 cwt. It was doubtless for this reason found necessary to provide the larger of the 2–6–2 tank engines of class '3'. These were of similar dimensions to the class '3' 2–6–0s already described and were fitted with the same boiler which was derived from that previously employed on the well-known Great Western engines of the 2–6–2 order, though shorter and fitted with a dome. The B.R. version turned the scale at just over 68 tons and had a maximum axle load of 16 tons 6 cwt (Plate No. 48).

The small 2–6–2s were a tank version of the class '2' 2–6–0 and call for no further description save that their total weight and maximum axle loads were 63¼ and 13¼ tons respectively. Why the latter figure should have been less than that for the corresponding class of tender engine the writer cannot say. But the information is derived from the published figures.

The Last Word

In writing the history of British locomotive practice we have often found ourselves in the fortunate position of being able to wind up a chapter or a section by drawing attention, as a climax, to some outstanding development. Thus our first period concluded with the emergence of the most typical of all British locomotives—the inside cylinder 0–6–0. At the end of the Victorian era we witnessed the breaking of new ground by the production of the 'Atlantic' and 4–6–0 types for passenger work. The Edwardian period is rendered noteworthy by the production of Britain's first 'Pacific', and that of the last years of the railway companies by the rival Gresley and Raven engines of the same type.

None of the developments surpasses, if any of them equals, that reserved for the conclusion of this final chapter. The B.R. 2–10–0 of class '9' (Plate No. 52) was the most powerful, non-articulated, series ever planned for ordinary use. The combination of two cylinders and a wide firebox placed over the last two coupled axles (though anticipated in the 'austerity' wartime goods engines) was a new departure for locomotives intended for all duties save the haulage of express trains.

The proportions and general plan of these remarkable locomotives call for more than passing note. The wide firebox was a virtual necessity for an engine of this capacity; no less was the employment of an adequate number of coupled axles to utilize the steam generated. The 2–10–0 type provided a good answer to these demands, but it involved certain restrictions, notably in the depth of the firebox and in the diameter of the boiler, whose upper limit was fixed by the loading gauge, while sufficient space had to be left beneath it for a firebox of adequate depth. The restrictions imposed by the general plan is also evident in the comparatively small coupled wheels—only 5 ft. in diameter—for the mixed duties envisaged. Nevertheless, these engines proved remarkably speedy in service, being reliably credited with several maxima of 90 miles an hour.

The limitations imposed upon the boiler proportions are reflected in the maximum diameter of only 6 ft. 1 in.—4½ in. less than that of the larger 'Pacifics'—and in the consequent reduction in the number of large tubes from 40 to 35. 138 small tubes were provided, only 2 fewer than on the 'Pacifics', but their diameter was reduced from 2⅛ in. to 2 in. Moreover, notwithstanding a greater length of 5½ in., compared with the class '7' 'Pacifics', the shallower firebox of class '9', placed as it was over 5 ft. coupled wheels, produced a heating surface of 179 sq. ft. and a grate area of 40·2 sq. ft., compared with 210 and 42 of the express engine. The 2–10–0 weighed 86¾ tons, of which 77½ were available for adhesion—a figure far in excess of that provided on any other class produced by British Railways. This excellent result was combined with a remarkably low maximum axle load of 15½ tons. To satisfy loading gauge restrictions the 20 in. × 28 in. cylinders had to be placed at virtually the same height as those of the 'Pacifics' and, of course, relatively higher compared with the axles. This entailed a steep slope which resulted in a somewhat ungainly appearance, and can hardly have been unattended by unfavourable effects due to the vertical element in the line through which power was transmitted. Nevertheless these engines were of remarkable conception and design and few would refuse to recognize them as representing a worthy culmination of the resource and ingenuity displayed by British locomotive engineers over the century and a quarter covered by the present volume.

INDEX

Index

Index

For Product Safety Concerns and Information please contact our EU representative GPSR@taylorandfrancis.com Taylor & Francis Verlag GmbH, Kaufingerstraße 24, 80331 München, Germany

Printed and bound by CPI Group (UK) Ltd, Croydon, CR0 4YY
01/05/2025
01858485-0001